AF516109

EVERY CREATURE HAS A STORY

Celebrating
30 Years of Publishing
in India

Praise for *Every Creature Has a Story*

'Lenin breaks down scientific information and research for a lay person in simple, easy-to-understand language ... Lenin makes even rats interesting.'

—*The Hindu*

'From looking at bird song to parenting behaviour and literary history to prey-predator relationships, the author takes on complex subjects with panache and offers scientific reasoning and research in each essay, separating fact from anecdotes ... It's bizarre, it's enthralling, and it's witty. And she does all of this with clarity and a scientific curiosity.'

—*Current Conservation*

'Hearteningly, Lenin attempts to tilt the scales by showing the centrality of creatures, great and small, in the natural order.'

—*The Telegraph*

'[*Every Creature Has a Story*] is illustrative of Lenin's own journey into the wild: a slow, gradual immersion into tales of discovery of the natural world and its ways.'

—*The Wire*

'Lenin, who is a veteran of the genre, writes with both amusement and adoration ... The book is a 101 of sorts, to the vastness and the wondrous inscrutability of the animal world.'

—*Outlook*

'A compilation of 50 essays on scientific revelations around behavioural patterns among a variety of animal species, Lenin's upcoming book aims at communicating the otherwise complicated phenomena in an easy, everyday language and at the same time making the narrative interesting enough for readers of all age groups and temperaments to engage.'

—*Firstpost*

EVERY CREATURE HAS A STORY

WHAT SCIENCE REVEALS ABOUT ANIMAL BEHAVIOUR

Janaki Lenin

HarperCollins *Publishers* India

First published in hardback in India by HarperCollins *Publishers* 2020
4th Floor, Tower A, Building No. 10, DLF Cyber City,
DLF Phase II, Gurugram, Haryana – 122002
www.harpercollins.co.in

This edition published by HarperCollins *Publishers* 2023

2 4 6 8 10 9 7 5 3 1

Copyright © Janaki Lenin 2020, 2023

P-ISBN: 978-93-5629-677-0
E-ISBN: 978-93-5357-707-0

The views and opinions expressed in this book are the author's own and the facts are as reported by her, and the publishers are not in any way liable for the same.

Janaki Lenin asserts the moral right to be identified as the author of this work.

Several chapters of this book have appeared in the *The Wire*. We thank them for allowing us to reproduce the articles.

All rights reserved. No part of this publication may be reproduced, stored in a retrieval system, or transmitted, in any form or by any means, electronic, mechanical, photocopying, recording or otherwise, without the prior permission of the publishers.

Typeset in 12.5/16 Dante MT Std at
Manipal Technologies Limited, Manipal

Printed and bound at
Replika Press Pvt. Ltd.

FSC
www.fsc.org
MIX
Paper from responsible sources
FSC® C016779

This book is produced from independently certified FSC® paper to ensure responsible forest management.

To all the creatures that inhabit Earth whose existence blesses me with wonder and joy.

Contents

Introduction

FOR three years, I revelled in my job of making science understandable to the readers of a new online publication, *The Wire*. I chose the best from a crop of ninety essays for this collection. The species ranged from spiders salivating during sex to snails entombing their parasites into their shells, from elephants being immune to cancer to giraffes solving the challenges posed by long necks. Before anyone points out I made a mistake in including a plant among these 'creatures', let me say there are no discrete categories in nature. It leaks, overflows, overlaps and intrudes across man-made boundaries. For this reason, I resisted grouping the essays in any way. There is a certain pleasure in delving into the disease immunity of bees and then leaping on to a fur seal beach in Antarctica. After all, the theme is the same: animals have found unique, and often bizarre, ways around difficulties facing their survival.

Mammals and birds staked out half the book, but their domination is not by design. I started every Monday not knowing what I would be working on that week. By the time I read the paper that made me gasp and the associated research papers, asked questions of scientists, and wrote it up, the deadline's alarm bells were screaming. In the intervening years since their publication, researchers found out more, and I have updated the articles to include the new knowledge.

The speed of the turnover wasn't ideal. As if the challenge of getting the science right and communicating it accurately weren't enough, I also aspired towards another goal – to entertain and connect with readers who knew nothing about the animals. I can't say I did justice to both every time, given the short deadline.

Putting this collection together allowed me to update and tinker with the stories until I could at least grudgingly say, 'This one is ready to fledge'. However, this doesn't guarantee I won't scowl when I re-read them a few months later. As long as I don't figuratively hurl them across the room, I am okay. Becoming better is a never-ending upward trajectory. I can't wait to reach my nineties, when I will be at the peak of my game.

This is a milestone along that ride.

Foreword

WE have, as human beings, long marvelled at the diverse forms of life around us. The kingfisher, cobra, banyan tree, grasslands, gaur, our domesticated pets, all bring completeness to our lives through a connectedness with nature. Charles Darwin had put it simply and beautifully in *The Origin of Species* (1861):

'Thus, from the war of nature, from famine and death, the most exalted object which we are capable of conceiving, namely, the production of the higher animals, directly follows. There is grandeur in this view of life, with its several powers, having been originally breathed into a few forms or into one; and that, whilst this planet has gone cycling on according to the fixed law of gravity, from so simple a beginning endless forms most beautiful and most wonderful have been, and are being, evolved.'

Humans marvel at the world and also try to understand it. This has, with each leap in our understanding, become an increasingly specialized enterprise. The earliest scientists were philosophers, poets, mendicants, priests, astronomers and astrologers. Early scientific explorations covered topics from astronomy, geology and our botanical and zoological diversity. These explorations, initially driven out of curiosity, found extraordinary and transformative applications in seafaring, commerce, agriculture and the growth of industry. As human societies grew and changed, science became the source of the river that fuelled technologies needed for commerce and conquest. Steadily, but surely, humans have moved from being just awed observers of nature to trying to understand nature's wonders, and now they have become nature's engineers and arbiters of the future of a planet we have brought to a precipice.

Thus, while endless forms most beautiful have evolved, we today have the power to both understand and to destroy them. Indeed, we have done both. As we realize this, it is all the more urgent to understand earth's many remaining natural wonders better, even as we strive to restore it to stability. It is this journey that Janaki Lenin takes us through in this delightful collection of essays.

Why does the nightingale sing? Is the song learnt or inherited? If the ability to sing is a proxy for fitness as a potential mate, how are the two linked? Did Moby Dick, the giant white whale, attack its tormentor's ship, the Pequod? Can the heads of sperm whales be used as battering rams or are they too sensitive? Humans were inspired by birds to want to fly, but did we learn how to break nuts from monkeys? Lenin takes a provocative question in each essay and analyses to answer it by peeling apart complexities and separating real possibilities from shaky hypotheses.

At the heart of the natural world is evolution. Life was 'originally breathed into a few forms or into one'. At one level, this means a shared chemistry for all life, linked through the thread of DNA. At another level, it means that extraordinary complexities and bewildering diversity in the behavioural repertoire of animals and in the interactions of animals and plants with each other. All of Lenin's essays, thus are important examples of how natural selection operates to display diverse behaviour.

In her essays, the author displays an ability found amongst the best science writers and communicators—the use of simple language, clarity of thought and rigour of logic—to communicate complex subjects without dumbing them down. Starting with a commonplace observation, or widely-held view, she hops across the analyses of the world's best specialists. Sometimes, we have an answer, sometimes we know what more we need to explore. Like a curious child prodding the parents relentlessly about almost every topic, Lenin takes us to the depth of a problem to tell us what lies at the core. That she herself was never trained as a scientist tells us both about her extraordinary talent and effort, and about how the keys to the doors of knowledge and understanding through science are there for all.

K. VijayRaghavan
Developmental Biologist,
The National Centre for Biological Sciences
Principal Scientific Adviser to the
Government of India

1

Good Singers Make the Best Dads[1]

An hour before summer dawn breaks, I lie awake listening to the song of drongos. At dusk, when I wind up work, oriental magpie robins sing. I would be honoured if birds sang for my enjoyment, but they sing for their supper, more or less. Males are the singers of the bird world, warbling to announce their strength and ownership of territory. Sometimes, I amuse myself imagining the birds are cursing their rivals, 'You cross-eyed, bow-legged buzzard. Step into my territory and I'll break your legs.' If only humans could scold as sweetly as birds.

Male nightingales (*Luscinia megarhynchos*) are the most famous songsters in Europe, ranking number one on the sophistication charts among forty-six bird species that scientists have investigated. These drably coloured bachelors have none of the visual flamboyance of other male birds like peacocks, but they more than make up for that disadvantage with their oral artistry.

After nightfall, male nightingales serenade their prospective mates from a repertoire of more than 180 songs, and hence the name – nightingale. Researchers categorize their extensive repertoire as whistle, buzz, and trill songs. Whistle songs attract females, trill songs deter rival males, and buzz songs highlight the singer's good health. They repeat their wide-bandwidth trills faster than any other bird, as many as 100 times a second. They can also sing 1,160 notes, more than any other bird studied. But not all bachelors are equally proficient.

One-year-old male nightingales have a handicap – their repertoire is less than half the size of older males.[2] The seniors are in peak form and press home their advantage. They leave their winter grounds in sub-Saharan Africa and are the first to arrive in Europe to sing and breed.

At midnight, female nightingales fly from male to male listening to their warbling[3], as if auditioning them, before settling for one. Couldn't this wait till daytime? Why do these little birds risk predation by owls to sing their hearts out? Birdsong rings out loud and clear on quiet nights, and the urge to breed seems to outweigh the risks.

Males invest so much effort and time in singing, they lose weight. After a few hours of sleep, they start singing again before dawn. Singing is not for the weak and lazy.

Among other singers, such as sedge warblers and European starlings, the females choose a mate who is a veritable jukebox with an extensive library of songs. But a male nightingale doesn't perform every number on his playlist. If he did, he would have to sing for about an hour, and unlike the judges of television talent shows, females won't listen patiently. Nor does he pick a random collection of melodies. Instead, he chooses his tunes and sings them in sequence depending on the context, time of day or night, and time of year. For instance, once a male finds a mate, he stops singing at night. When he warbles during the day, he skips the flirtatious whistle songs.

Nightingale songs don't only say – 'I'm a King Bee, Baby', 'Stay out of my territory', and 'Won't you be my love'. There's more to these songs, say Conny Bartsch and her research team at the Freie Universität, Berlin, Germany. They studied twenty male nightingales in a wetland outside Potsdam, and they say birdsong advertises how good a father the singer is.

Nightingales, like males of about 80 per cent of bird species, are hands-on partners and fathers. For two weeks, they fetch food for their mates who are incubating eggs. Once the eggs hatch, they feed the chicks. They also protect their nests from the predatory tawny owls. Often, before the young become independent, females may lay another clutch of eggs. The fathers work hard to feed their mates and the growing chicks. If they shirk their duties, their mates struggle to pick up the slack, and their young pay the price.

When the future of the next generation is at stake, how is a female to know if the prospective mate would make a good father? Could she use his singing stamina to gauge his commitment to the relationship? Bartsch and her colleagues say there is an association between singing and hunting.

Discerning females seek mates who sing the same sequence of songs in a loop that includes many low-frequency buzzes and high-pitched whistles and trills. 'Although we cannot say what exact song features female nightingales finally use to assess males, it seems most likely that females use both the quantity (repertoires) and quality (performance) of songs to assess a male's adequacy as a mating partner,' write Bartsch and her colleagues.

Once the chicks hatch, these ardent singers visit the nest on an average eleven times an hour bearing insects and worms, investing the same amount of effort as the mothers in child-rearing.

Since a male nightingale invests heavily in reproducing, it pays to advertise his fathering skills up front. Apparently, the song that makes female nightingales weak-kneed goes something like this:

'I will be a great father
Have a little faith in me
I will be a wonderful husband
Just watch me rear our babies.'

The musical virtuosos, the older experienced males, know the lay of the land and are better at finding food for their offspring. It's no wonder then that they are the most sought-after partners for the season.

The bar for skilled performance is set so high, up to half the males don't find partners.[4] Their songs of courtship and self-advertisement ring melodiously through the nights until late July when it's time to leave for Africa.

When these young male nightingales return the following year, older and wiser, their songs would have gained in intricacy, and they may even win mates.

2

Chimpanzees in Mourning[1]

ARE primates capable of grieving? Do they know what death is? An animal would have to recognize the irreversibility of death to grieve for a dead member of its group or family.

Many female primates, from bonnet macaques to gorillas, carry the corpses of their dead infants for weeks. Sometimes, the bodies reach a state of mummification before the mothers abandon them.

In 2011, Peter Fashing of California State University, Fullerton, US, and his colleagues reviewed[2] fourteen instances of gelada mothers carrying the corpses of their infants. Geladas, close relatives of baboons, live in the Ethiopian highlands. Not only mothers but also juvenile females groomed and hauled the dead bodies. The other members of the group didn't avoid the ones carrying the foul-smelling corpses and instead tolerated the stench. After they stop lactating and restart the ovarian cycle, most such mothers abandon the remains. But one gelada mother continued to hold the mummified corpse of her offspring even while mating. The authors surmise two possible reasons for this behaviour – the delayed decomposition of bodies because of cold conditions may prolong the geladas' interest in them, or such reactions to death may originate in ancient primates.

Many species of primates are social animals, eating, playing, grooming and sleeping together. Within a group, individuals form bonds, spending more time with each other. How would the death of an adult affect the others? While there are many instances of mothers hugging the remains of their infants, primatologists have much fewer opportunities to study a group's response to the death of an adult.

Edwin van Leeuwen of St Andrews University, Scotland, and his colleagues published one of the few instances of a chimpanzee group's reaction to the death of a member.

Within the 4,200-hectare Chimfunshi Wildlife Orphanage Trust, Zambia, 120 chimpanzees live in six groups enclosed in large outdoor enclosures. Although many of the primates were rescued from illegal trade, most of the forty-three animals in 'Group 2' were born in captivity. Thomas, a nine-year-old chimpanzee, had been born to a wild-caught mother. His group roamed through a sixty-five-hectare fenced forest.

In May 2010, Thomas had been missing for two days before he was discovered lying dead. Earlier, the keepers had walked past that spot and not seen the body. They concluded the primate had died elsewhere, and some chimpanzees had dragged his body to the site. The necropsy report said he died of pneumonia.

According to the staff, Thomas had been a social animal, parking himself by the door of the indoor holding facility.

He and his companion Pan, another male chimpanzee, called, greeted and played with any passing chimp. They paid particular attention to adult females.

With a video camera, the keepers recorded how the rest of the group reacted. The chimpanzees had splintered into subgroups to forage in different areas. Some members gathered around Thomas's body while the rest may have been elsewhere, unaware of his death.

Within minutes of the body's discovery, Pan, the dead chimp's buddy, picked up a tree branch and lunged towards the body, sending the others screaming and scurrying out of the way. When he hit a female, or at least hit the ground near her, she chased him into the forest. He returned, forcing his way close to the body every few minutes to examine it once again. About fifteen minutes later, he pulled tree branches as he ran towards Thomas's body. He disappeared from the scene while chasing a female. Pan may have been distressed and frightened by his companion's strangely inert form. Or his aggressive display may have been his way of grieving.

Ignoring the caretakers' calls to feed, Violet, the dominant female, who had been watching Pan, stood up on her hind legs and screamed. She rocked back and forth as she made her way towards the body. When she was close enough, she hit it hard before running away.

Except for these dramatic interludes, most of the others gathered quietly around the body, while some inspected it. Van Leeuwen says sitting still is not typical chimpanzee behaviour.

After Violet left, Noel, another adult female, approached the corpse and cleaned its teeth with a grass stem. Chimps examine and clean each other's teeth. But this is the first report of one cleaning the teeth of a dead one and using a tool for it.[3] She then stuck the stem into her own mouth. Noel wasn't interested in what he had eaten since she ignored the keepers' call to feed. Picking Thomas's teeth seemed to be meaningful to her, says van Leeuwen.

Thomas had been especially close to the thirty-three-year-old Noel, who had adopted him after his mother died four years earlier. He had groomed and played with Noel and her two offspring. While she poked around his mouth, her teenage daughter, Nina, sat with her and silently observed her mother.

Chimp reactions, ranging from aggressive displays to grooming, may be their way of making the inert body react.[4] Or perhaps the animals were disturbed, recognizing the permanence of death. Sometimes, older individuals in a group may even guard the corpse from other members.[5]

James Anderson, of Kyoto University, Japan, highlights[6] the actions of chimpanzees in Taï National Park, Côte d'Ivoire. They groom and lick the wounds of their injured friends but don't take the effort with the dead, suggesting they know a cold, stiff body has no life in it.

Van Leeuwen says Thomas's case is the first account of chimps responding to the death of an adult group member. Compared to the deaths of infants, many more adults attended to the body of the grown chimp, and their reactions were much more elaborate.

When two infants of the same group died earlier, the rest didn't seem to react. Pan, for instance, didn't respond in the same manner as when Thomas died. The researchers speculate that the death of an adult may trigger more visceral responses than a youngster's death. 'I think the reactions of Noel and Pan (in comparison to the other chimps) show that in chimps, the bonds in life determine the responses to death,' says van Leeuwen.

When an elderly captive female at Blair Drummond Safari Park in Scotland[7] was nearing death, her three group mates groomed her until she died. The two females sat in silence, but the young male, like Pan, attacked the corpse. The researchers thought he may have been trying to resuscitate her. The daughter refused to leave her dead mother's side as if holding a vigil. For the following five nights, the group refused to sleep in that room or touch anything associated with the dead chimpanzee.

While groups eventually accept the loss of a member, some may not. Jane Goodall recorded one young chimpanzee 'sank deeper into depression' after the death of his mother. A few days later, he curled up and died at the nest he had shared with her.[8]

Van Leeuwen points out that the female chimpanzees sat closest to the body while the adult males sat farther away. 'This was no surprise for the Zambian chimpanzee keepers as their mourning ritual is very similar. They told us that women stay close to the body, in silence, and the men sit outside the house. Since Thomas was well-integrated into the group, perhaps his death upset the other members.'

The primates may not be all that different from us in recognizing and grieving for the dead, even though we don't hit or manhandle corpses.

3

The Eye of the Chameleon[1]

It's impossible to sneak up behind a chameleon. Without taking an eye off a juicy grasshopper, it can swivel the other eye around to watch you. With each eye pinned on different objects, does the chameleon know what each one is watching?

An ambush predator, the chameleon is famous for its ability to change colour. It conceals itself in trees and freezes in position indefinitely, waiting for insects to come within reach of its long, sticky tongue. If it has to move its head or adjust its body position to look in another direction, its cover may be blown. Its prey would fly away, or a raptor may snatch it up. Independently moving eyes is a splendid trick to overcome this problem. Some birds, fish, and seahorses also have the same talent. Each eye is controlled by the opposite side of the brain so the brain's left hemisphere knows what the right eye is doing, and the right hemisphere the left.

Chameleon eyes are the coolest among vertebrates in more ways than one. The eyelids are fused to the eyeballs, leaving a tiny hole for the pupils. The lens within the eye makes the chameleon near-sighted, but the lizard can spot prey five to ten metres away without needing spectacles. A unique combination of a minus-powered lens with a far-sighted cornea gives the chameleon a telephoto eye.[2] A small insect must appear as big as a crow to a chameleon.

Its domed eyes bulge outwards, so the flat, thin body doesn't block the view. Scientists say the range of vision is more than 180 degrees horizontally and 90 degrees vertically on each side. This gives the chameleon an almost unrestricted view in any direction without moving its head. In comparison, humans see 80 degrees across and 70 degrees vertically.

Hadas Ketter Katz and Gadi Katzir, along with colleagues in the University of Haifa, Israel, tested whether there's any coordination between a chameleon's eyes. The most accurate way to test sight abilities is by using computerized images, so the experimenter has control over size, speed, contrast and distance.

It took Katz a whole year to build the experimental apparatus in a way that chameleons would be comfortable enough to perform the experiment voluntarily. She didn't use restraints but lured them out of their enclosures with a live cockroach. She rewarded them randomly between tests, so their performance was not influenced by treats.

Katz trained thirteen common chameleons (*Chamaeleo chamaeleon*) found around the Mediterranean to focus on a moving target on a computer screen. Many creatures, like toads, praying mantises, and archer fish, try to catch virtual insects.

When there was only a single object to track, chameleons focussed both eyes on it. The researcher's black rectangle with two little dangling legs apparently resembled an insect closely enough that the chameleons leaned forward and smacked it with their long tongues.

This is where Katz added a complication to the experiment. She split the black rectangle in two, and each travelled across the screen in opposite directions. Researchers had not conducted such an experiment before with chameleons.

As if they had seen this strange self-replicating, confounding insect all their lives, the slow-moving lizards weren't fazed by it at all.

'I had no idea whether the chameleons would binocularly track one of the targets, or if they would split their gaze and track each of them monocularly as they did,' says Katz.

Each chameleon continued to watch the two parts of the insect drifting away from each other until it decided which black rectangle it liked better. Once it made that decision, the other eye swung around to track the favoured target along with the first eye. The authors say this demonstrates that each eye knew what the other was watching and did not function independently. Katz was excited by the lizards' performance. She confesses, 'Just imagine me all alone in the lab screaming and jumping.'

If humans had this remarkable ability, we could watch a football match and a movie at the same time and bring our eyes together for the goal or climax.

To gauge distance, we need to triangulate with both eyes. But each chameleon eye is fully capable of estimating distance on its own.[3] Scientists taped one eye shut, and the chameleons shot out their tongues to snag insects accurately. If they don't need to use both eyes to assess how far away their prey is, why do they focus them on the target before sticking out their tongues?

Jack Pettigrew of the University of Queensland, Australia explains why there would be a problem. 'Prey capture is only possible when both eyes are looking in the same direction with the same plane of focus, at the same object, which minimizes the chance that prey capture will be directed at two different targets.'

Chameleons have the advantage of eyeballing two separate objects with their brains controlling each eye separately. But once they decide which insect to eat, they impose a higher level of control and lock both eyes on the chosen target before shooting out their tongues.

If chameleons had truly independent eyes, their brains couldn't handle the disparate inputs.

4
The Oddest Bird in the World[1]

THE chicken-sized kiwi may look odd, but it evolved to make the most of its nocturnal lifestyle. It scrounges along the forest floor while tapping its long beak, much as a visually impaired person taps a cane. Unlike other birds, its nostrils are at the tip of its bill. But the nose-end does much more than smell. It is covered in sensory pits, called mechanoreceptors,[2] that feel earthworms and grubs wriggling underground. This sensitive bill-tip is the kiwi's primary sense organ, and the bill is its main tool. It probes the ground down to twelve centimetres in depth, as an ibis would in water, and sniffs loudly.

The kiwi uses its sense of smell not only to find prey but also to communicate. It gives off an unusual odour that Sir Walter Buller, a nineteenth-century naturalist, described as a 'peculiar earthy smell' in his book *A History of the Birds of New Zealand*.

The bird evacuates its pungent-smelling excreta at prominent locations, such as logs and roots, to mark its territory. Should a neighbouring kiwi come upon this malodorous dropping, it waves its beak in the air and snorts[3] before going its way.

Despite the many peculiar features, it is the size of the dumpy bird's eyes that foxed biologists. The eyes are very small relative to the body mass. Unlike other flightless birds like ostriches that have large peepers, the kiwi's tiny eyes cannot see the end of its own long bill. Its field of vision is the poorest among birds. The optic nerve is thin, and although the brain is large for its body size, the lobes that process visual information are very small.[4]

However, other night-time birds such as owls have large eyes to see in the dark as they fly. Since the kiwi can't get airborne with its nubby wings, it doesn't have much use for vision. By trading sharp vision for acute hearing, touch and smell, the smallest of the flightless birds behaves more in the manner of small nocturnal rodents.

A group of fifteen molecular biologists from six institutions in four countries collaborated on a four-year project to sequence the kiwi's genome, the largest bird genome known to date, to study which genetic alterations make these unique adaptations possible. The researchers extracted DNA from the embryos of two eggs of the North Island brown kiwi (*Apteryx mantelli*), the most common and widespread of the five kiwi species that are found only in New Zealand.

Eyes need a lot of energy and large ones need even more of it. Since the kiwi has the lowest metabolism of any bird, the researchers suggest that may limit the size of its eyes.

'I do not think it will hold up since many birds with much higher metabolic rates than kiwi have small eye sizes,' says

Graham Martin, of the University of Birmingham, UK, who specializes in sensory perception of birds. Instead, the kiwi's eyes became smaller as it has no need for powerful vision. It didn't have to watch out for predators throughout its evolutionary history.[5]

The scientists found the green and blue colour receptors in the bird's eye were non-functional, leaving the kiwi without colour vision. But it had more smell receptors than any other bird, enabling it to detect a wide range of odours.

The kiwi wasn't always a denizen of the night. Millions of years ago, on its arrival in New Zealand, its ancestor hunted by day. But by then, the now-extinct, three-metre-tall moa had already established itself as a daytime feeder. The researchers say the kiwi was forced to become a creature of the night to survive. They peg these adaptations to have occurred around thirty-five million years ago.

Martin says it is not yet clear if competition with the moa forced the kiwi to adapt. Or if the bird's nocturnal habit of foraging on the forest floor influenced the changes. 'It could just be that no other animals were exploiting this rich food source.'

In the study, the biologists don't mention the genetic origin of the mechanoreceptors on the bill tip.

'It would be impossible to look at the mechanoreceptors in the bill with the data we have,' says Diana LeDuc, the main researcher. To examine these sensory pits, it is not enough to study the genetic code alone. The scientists would need to sequence the transcriptome, or RNA molecules that translate the genetic code into proteins, of the mucus inside the bill to see which genes are active.

Despite the uniqueness of the kiwi's beak, it has one drawback. While the bird pokes its beak into the ground, the nostrils get plugged up, and the kiwi snorts loudly. Since it evolved on the predator-free islands of New Zealand, there was no need to be silent. However, these earthbound birds now have to contend with legions of dogs, cats and weasels, brought by colonizing Europeans. Although there are about 25,000 brown kiwis in North Island, the numbers have been falling for decades.

Since only 5 per cent of the chicks make it to adulthood, New Zealand's Department of Conservation runs an artificial incubation programme called 'Operation Nest Egg'. For a chicken-sized bird, its 450-gram eggs are enormous and incubated by males. In the standard scale of body weight to egg size ratio, these huge eggs ought to be produced by a thirteen-kilogram bird like an emu.[6] The project takes these eggs from wild nests in underground burrows or hollow logs and incubates them artificially for about eighty days. Chicks are reared until they reach 850 grams in weight before being released in the wild. At that size, at least the stoats won't get them.

Deadly predators aren't the only threats. The kiwi's genome itself may work against it. Despite its size, it doesn't vary much between individuals, Diana LeDuc says. How well a species adapts to changing environmental conditions and its long-term survival depends on having as many differences as possible. The diversity of the kiwi genome was as low as inbred birds, and further insights from the genome could aid conservation management planning.

'Low genetic diversity and inbreeding lead to further population decline,' she says. 'Hence, making use of the genome, the diversity of that population can be determined, and an informed decision can be made about how the different kiwi populations should be mixed.'

The world would be a poorer place if such an enigmatic, unique bird were to disappear.

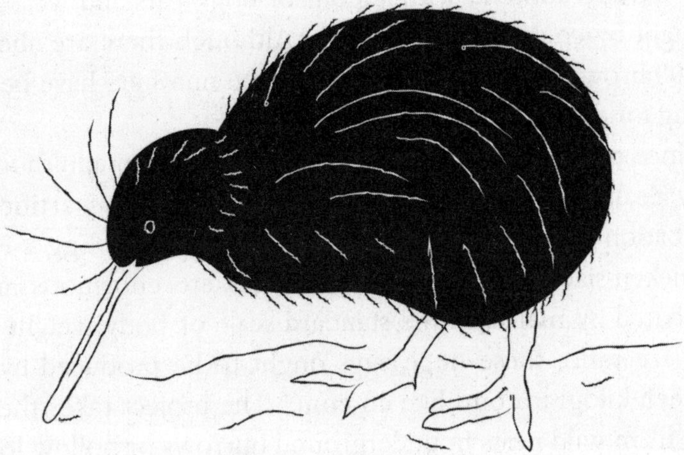

5
Vaccinated Bees[1]

WHEN bacteria, virus, or other pathogens attack our bodies, we produce vast quantities of antibodies. Secreted by blood, these proteins stick to the invaders and aid their destruction. This is our primary defence mechanism. Insects, however, lack antibodies, but they don't keel over and die at the first symptom of a disease.

Social insects such as honeybees live in densely packed hives and ought to be prone to infection. Queen bees rarely, if ever, leave the colony. Hard-working worker bees range far and wide to gather pollen and nectar. While they go about their business, they pick up germs. Back at the hive, they use the contaminated pollen to create royal jelly to feed the queen. These pathogens could blaze through the nest and destroy it. How do insects survive disease?

Scientists thought if insects didn't have antibodies, they must have a primitive immune system. Crystal cells in the hemolymph (or insect blood) immobilize the invading microbes, and plasmocytes (like our white blood cells) swallow them. But that's not the only defence.[2]

After the researchers injected bacteria into fruit flies, the hemolymph ignited with antibacterial activity. A fat body (similar to our liver) within the circulatory system produced quantities of infection-fighting peptides[3]. Even dead bacteria triggered this reaction.

Some insects, like mealworm beetles, red flour beetles, and cabbage leafhoppers, don't stop with putting up a fight against pathogens. They confer the resistance they develop to their brood, similar to children inheriting antibodies from the mother. What is the mechanism of this antibody-free immunity?

Scientists led by Dalial Freitak, a postdoctoral researcher at the University of Helsinki, found the key is vitellogenin, a hormone-like protein found in egg yolk and the hemolymph of insects.

Within the egg, it is a nutrient for developing embryos.[4] But it is vital in healing inflammation and injury. It also dictates lifespan – the hemolymph of short-lived worker bees has concentrations of up to 40 per cent of vitellogenin while longer-lived queens have up to 70 per cent.

Freitak met Heli Salmela, who was working on vitellogenin for her doctoral thesis, at a group meeting at Helsinki University, Finland. 'She presented her data on the capacity of vitellogenin to bind with bacterial pieces,' says Freitak.

Since it is also an egg protein, could vitellogenin transmit immunity to the next generation? Together with Gro Amdam, at the Arizona State University's School of Life Sciences, they designed an experiment to test that theory.

They applied vitellogenin to some dissected ovaries of queen bees but not others, and then introduced fluorescent fragments of *Escherichia coli* to all of them. The ones that didn't get the protein didn't absorb the microbial pieces. When the researchers saw the vitellogenin-treated ovaries sucking up the fragments, they grew excited.

'I felt like I had won the lottery!' says Freitak. 'I don't think I have ever been as happy as when I saw the images of ovaries, full with fluorescent bacterial pieces in the vitellogenin treatment. We danced in the lab with Heli.'

The researchers also proved vitellogenin, and not any of the other proteins in the hemolymph, played the leading role in triggering an immune response.

'What we found is that it's as simple as eating,' Gro Amdam said in a statement.

When honeybee workers bring pathogens back to the hive and feed the contaminated jelly to the queen, she digests and stores them in the fat body. Her hemolymph carries the protein-wrapped microbial strands to the developing eggs in the ovaries, priming the immunity of the next generation. Bees are born vaccinated and ready to fight diseases in their environment. These insects had evolved a vaccination system millions of years ago. The mystery of how insects inherit immunity that had vexed scientists for a long time had been solved.

Vitellogenin bonded more strongly with gram-positive bacteria like American and European foulbrood diseases than with gram-negative bacteria like *E.coli*. Spores of the bacteria that causes American foulbrood disease spread quickly through hives, wiping out entire colonies. The researchers suggest the intense reaction to these gram-positive microbes may reflect the severity of threat they pose to honey bee larvae.

This inherited immunity probably doesn't last long in the bloodline and may be replaced by specific antidotes to other diseases as they occur in the bees' environment.

Bees and other pollinating insects are crucial for agriculture, fertilizing the flowers of fruit and nut trees and vegetable crops. The United Nation's Food and Agriculture Organization estimates that 90 per cent of food in 146 countries is supplied by more than 100 crop species, most of which are pollinated by insects.

Now that the researchers have figured out how this defence mechanism works, they can develop an edible vaccine to deadly diseases like the foulbrood and feed it to honeybees. Conversely, they can also interfere with this acquired defence to control pest insects and their damage to crops.

Freitak says her team is testing and patenting a harmless vaccine against American foulbrood disease that can be introduced into bee hives in an edible cocktail. 'They would then be able to stave off disease,' she says.

This would not only shore up honeybee immunity but also reduce the threat to human food security.

6

The Scent of a Fur Seal[1]

EVERY year, millions of adult Antarctic fur seals (*Arctocephalus gazella*) heave themselves on to the beaches of South Georgia Island, about 1,400 kilometres southeast of the Falklands, in the densest gathering of marine mammals in the world.

Seal cows give birth to a single pup between mid-November and late December and nurse their young for approximately four months. The forty-kilogram mothers cannot stay with their offspring through this time. They dive into the sea, travelling as far as 240 km for five to ten days in search of food. Once they have eaten their fill and are heavy with milk, they waddle ashore to a beach crowded with youngsters, nursing mothers and bulls. If they cannot find their own pups, the young can die because they are unlikely to be adopted by other moms.

Antarctic fur seals are good at locating their offspring, only 10 per cent of 832 mothers adopted another pup.[2] Comparatively, 90 per cent of Hawaiian monk seals nurse young that are not their own. How do the fur seals find their progeny?

Could the memory of where they parked their offspring guide them to their babies? The youngsters don't sit still. With their mothers gone, they get together and roam the beach, playing and jostling. [3]

For many years, biologists thought the seals find each other by their calls.[4] Amongst the hundreds of raucous animals, the hungry babies seem to miraculously tune into their moms' voices and approach them. While frequency modulation carries for sixty-four metres, amplitude doesn't travel more than eight metres.[5] Perhaps, they hear the frequency modulation first, and when they get close, they rely on the amplitude. When mothers and pups meet, they touch noses, sniffing each other vigorously. Did smell play a role in recognition?

Amphibious mammals, like seals, can't sniff underwater, where they spend most of their time. They filter big gulps of water for tiny shrimp-like krill and hunt fish by sight. If the animals don't rely on their noses much, could the sense have atrophied?

In a 2011 study, researchers led by Benjamin Pitcher of Macquarie University, Australia, offered stuffed toys smeared with the smell of their own and unfamiliar pups to Australian sea lion cows. The mothers preferred the ones that smelt like their young. Since the mute toys were identical, the only cue the cows had was the smell.[6]

Researchers from Bielefeld University, Germany, and The British Antarctic Survey tested if smell plays a role in confirming the identity of fur seal parent and offspring.

They analysed the skin secretions from forty-one mother fur seals and their pups from two different rookeries on Bird Island Research Station, adjacent to South Georgia Island. A unique composition made up of about thirty-five substances coats every animal's body. In most other mammals, the chemical signature changes with age and readiness to mate. Fur seals are likely to be no different. Despite these alterations, mothers and pups smell similar to each other sharing at least twelve substances. The young may inherit their chemical signatures from their mothers, and they may also pick up some compounds from the beach where they grow up.

In a process that scientists don't understand yet, moms and pups reunite at the same spot where they were together last.[7] The mothers nurse their babies for up to three days, and then they wrench themselves from their fast-growing offspring and go out to feed in the sea again.

Although biologists have long known animals can smell blood relatives, Barbara Caspers, one of the researchers, says, 'this has not been chemically proven until now'. Since sense of smell is the oldest and the most widespread among animals, 'there is reason to believe the mechanisms are similar across the animal kingdom and not just fur seals.'

Indeed, we humans are no exception. Babies recognize the odour of their mother's breasts,[8] and mothers, in turn, recognize the smell of their babies.[9] Can mothers pick out their own from a roomful of bawling infants? Usually yes, but this is not a specialty of mothers alone. When fathers spend as much time with their offspring, they can identify their children by smell too.[10] But humans don't run the gauntlet of being separated from their infants and having to retrieve them from a crowd.

Fur seals' sense of smell doesn't only lead to mom-baby reunions. It has another equally critical role – females follow their noses to find unrelated mates.

During the eighteenth and nineteenth centuries, Antarctic fur seals were hunted to the point of extinction. Only a small colony survived on Bird Island. With protection, this colony grew and colonized neighbouring islands. Although there are millions of Antarctic fur seals now, they descended from a small founding stock. Inbreeding could be a significant problem. This is where smell could be the species' saviour.

The more diverse the ancestry of the flap-footed mammals, the more complex their scent profile. In other words, if their ancestors come from a vast gene pool, they have many more substances coating their skin. Such seals would be sought-after as mates since they stand a better chance of survival in a species that went through a genetic bottleneck in recent years.

Like many other seal species, aggressive Antarctic fur seal bulls smell rank during the breeding season. Apparently, cow seals have a body odour fetish. Since every animal's signature scent is genetically coded, fur seals can tell blood relatives from strangers, preventing inbreeding.

At the colony level too, scent substances could play a role. Seals may not show their aggressive side to kin, thus forming stable relationships, says Pitcher. 'We now need to present these chemical cues to seals in behavioural experiments to determine if they can use them for recognition and which cues are important.'

7

Conservative Chimpanzees[1]

CHIMPANZEES have culture. By culture, biologists don't mean the primates sing and dance. They say an animal has culture when it learns to do something by watching others.

Some populations of the primates dip a stick into a trail or nest of ants and wait for the insects to swarm over it. In Guinea, chimps use long sticks to poke into nests and short sticks to pick up ants running along trails. Once ants swarm over the twig, the primates swipe a hand along the length of the tool and stuff the biting mass into their mouths. Or, they nibble the ants straight off the stick, as we would eat a skewered sheesh kebab.

Kathelijn Koops of University of Zurich, Switzerland, and her colleagues from Germany and Japan compared neighbouring groups of eastern chimpanzees (*Pan troglodytes schweinfurthii*) living in the same habitat in Kalinzu forest, Uganda. By studying genetically similar groups living in the same area, the researchers could rule out any behaviour that can be explained by innate predisposition as well as behaviour arising out of opportunities inherent in the environment. One group was called 'M', after the musanga trees that were common in its territory. The other was 'S', or Sunzu, the name of the area. The members of this group were so unused to humans that the 'S' may as well have stood for 'shy'.

The researchers found the sticks used by the M-group were on average 16 centimetres longer than the ones used by the S-group. Since the M-group chimps were habituated, the researchers could watch them fishing for ants. The primates used a clever technique to avoid being bitten by the formidable soldiers of the ant colony. They hung suspended from an overhanging tree when sticking a long twig into the ant nest. Perhaps, fishing while hanging from trees required longer sticks to reach deep into the nest. What's more, the length of the ant-fishing stick has remained constant since 1997.

The researchers could not observe the shy S-group at dinner. They could only measure the length of the sticks they found lying abandoned near ant nests in that group's territory. These chimps used shorter sticks to fish for the same species of ant. In fact, these stems were far shorter than those made by any other chimp groups.

None of the nests that the S-group raided had a tree nearby, so it's unlikely they dangled from one while poking into the nest. The researchers say it is possible this group nibbles ants directly from the stick, another reason the chimpanzees may not need a lengthy probe. They also speculate that there may be differences in the structure of the ant nests in the territories of the two groups. Perhaps, the ant nests were not deep, or the ants not aggressive enough. This is the first study to show cultural differences in tool manufacturing between neighbouring groups.

'One possible criticism [of the study] could be that the data for S-group are relatively limited and the difference in tool length could potentially reflect a different prey preference in this community,' admits Kathelijne Koops. 'However, even if S-group differs in army ant species preference, this would still mean that there is a cultural difference between the two communities because prey availabilities are identical.'

In chimp society, there is little social interaction between adjacent groups, making it difficult for one group to learn from its neighbours. On reaching sexual maturity, males usually stay home, while females leave to join other groups. No matter what their natal group did, once these immigrants join a new group, they adopt its behaviour. Eventually, there's no difference in ant-fishing techniques between females and males of the same group, offering more evidence of culture among these primates.

Primatologists had already established chimpanzee culture in the Côte d'Ivoire in 2012. Lydia Luncz and her colleagues from the Max Planck Institute of Evolutionary Anthropology, Germany, studied the differences between three neighbouring groups of another subspecies, the western chimpanzee (*Pan troglodytes verus*). Each group had its own way of doing things, such as how it used tools, what it ate, how it hunted, and how its members interacted with one another. The researchers totalled up twenty-seven activities in which no two groups used the same technique.

What is special about this study?

'Finding such sophisticated cultural diversity in another subspecies of chimpanzees, besides the western chimpanzee, means that this ability of creating and maintaining differences between neighbours already existed before the split of chimpanzee subspecies,' says Luncz. 'If this behaviour is found to be universal in all chimpanzee communities, the likelihood increases that it was already shared with our last common ancestor about five to seven million years ago.'

However, not all eastern chimpanzees use sticks or eat ants. 'These findings in Uganda, where I also work, make me wonder why our Budongo chimps, a few hundred kilometres away, are so limited in their own toolset,' says Thibaud Gruber, from the University of Neuchâtel, Switzerland. 'They do not display any ant-fishing, or stick use, for that matter. One hypothesis is that the Budongo communities lost stick use for an ecological reason. Our chimps do not feed on ants at all, and on hardly any termites. Ants are just not part of their diet. Actually, in this respect, Kalinzu is unique in Uganda.' Perhaps their lack of interest in these insects makes the use of stick cutlery unnecessary.

Despite their remarkable ability to learn by watching others, the insular nature of chimpanzee society, not having social contact with other groups, slows the spread of cultural innovation.

8

Sex Change in Bearded Dragons[1]

SOME female bearded dragons of Australia are actually males. They are not in drag; they are fully functional females that mate with males. These dragons even beat regular females at their own game by laying many more eggs. The expert sex-reassignment surgeon is temperature.

In humans, women have XX sex chromosomes, and men, XY chromosomes. Sperm dictates the gender of the foetus because the Y chromosome has the SRY gene that governs male development. It generally does an even-handed job, creating an equal number of boys and girls.

Lizards have sex chromosomes too, but that is where the similarity with humans ends. Instead of having two of the same kind of sex chromosome, female lizards have ZW chromosomes and males have ZZ chromosomes. In this reptilian version, it is not the sperm that decides the gender of the offspring, but the egg.

It gets freakier in reptiles like crocodiles and turtles that have dispensed with sex chromosomes entirely. The sex of their babies is not determined when sperm meets egg, but in the middle of the incubation period, a few weeks after conception. The temperature at a particular phase in the incubation period decides whether the embryos develop testes or ovaries. Since heat affects the whole clutch, the sex ratio is often skewed in favour of one gender.

For a long time, biologists thought the gender of reptilian offspring was governed either by genes or temperature, but not both in the same species. In 2007,[2] biologists discovered the central bearded dragon (*Pogona vitticeps*) doesn't strictly follow one system. Although chromosomes assign the gender of embryos at the time of mating, heat overrides genes and turns genetic males to females.

While these sex-reversed lizards were created in labs, this is the first time that biologists report finding trans-female bearded dragons in the wild. This is the equivalent of women with XY chromosomes, called Swyer syndrome. While they cannot conceive without medical assistance, the lizards, however, are fertile.

The central bearded dragon, found in the dry interior of Australia, is, so far, the only mammal, bird, or reptile known to perform such a gender switch. Over the years, Clare Holleley, Arthur Georges and their colleagues from the University of Canberra and La Trobe University, Australia, found females with male genotype in the wild. Of 131 lizards they caught and analysed, eleven turned out to be females with ovaries and male ZZ chromosomes. Most of these males-turned-females occur along the border between Queensland and New South Wales, over an area of 23,650 square kilometres

Arthur Georges, senior author of the paper, says when he and his team first discovered sex-reversed females in the wild, they 'were pretty excited because it opens up the possibility of ZZ genotypes reversed to females mating with normal ZZ genotypes [males] to produce viable offspring in the absence of the W chromosome. Temperature takes over the role of ensuring both males and females are produced.' He calls the discovery stunning, like finding humans giving birth to boys and girls without the Y chromosome.

Holleley and her team bred some of these trans-females with males in the lab and compared their reproductive success with regular ZW females. The sex-reversed females won hands down, laying twice as many eggs as regular females.

The ZZ-chromosome mothers bred with ZZ fathers and made ZZ offspring. There was no chance for a W chromosome to sneak in. Within one generation, at least in the lab, there were no genetically female lizards in the sex-reversed lineage.

The biologists artificially incubated 389 eggs from twenty-one clutches at various temperatures. When eggs of sex-reversed mothers incubated at low temperatures, between 26°C and 28°C, the hatchlings were all male. At the critical temperature of 33.5°C, chromosomally male embryos started developing ovaries. A rise of one degree on the thermostat turned the entire clutch to female. Sex chromosomes had become redundant.

In the offspring of regular mothers, sex chromosomes determined gender between 22°C and 32°C, and the sex ratio was even. When the temperature shot up above that threshold, it interfered with the role of sex chromosomes. Male embryos of ZW mothers changed gender at the critical temperature of 34.7°C. At 36°C, the offspring were all females.

In short, hatchlings of genetically female mothers waited until the temperature rose higher before switching gender. However, the offspring of genetically male mothers inherited temperature sensitivity, changing sex at a lower temperature. If heat didn't play a major role, ZZ mothers would produce only sons.

'I think it's an elegant piece of work that drives a final nail in the coffin of an older paradigm [either genes or temperature regulating sex but not both in the same species],' says Rick Shine, of the University of Sydney. 'Clearly, genetically determined sex and environmentally determined sex are not poles apart – a species can switch from one system to the other in a single generation.'

While this ability to switch from one system to another sounds cool, it may not be a boon for the species. The impending spectre of rising temperatures caused by climate change could hit bearded dragons severely.

'Initially, as climates warm, this species will move from genetic sex determination to temperature-dependent sex determination,' says Georges says. 'All will be well provided temperatures range in such a way as to produce both males and females. But if the climate continues to warm, then there is a risk that only females will be produced, leading to male limitation and population crash.'

Until now, reptile biologists and conservationists were worried how crocodiles and turtles that use only heat to set the gender of their embryos would fare under global warming. Now it appears even reptiles with sex chromosomes may be affected.

9

Wasps Enslave Spiders to Weave for Them[1]

MANY wasps are parasites. They paralyse caterpillars or spiders by stinging, lay their eggs in them, and seal them alive inside their nests as ready-to-eat meals for their larvae once they hatch. If that isn't diabolical enough, some wasps take victimization to a whole new level by becoming zombie masters.

Reclinervellus nielseni, a wasp known only by its scientific name, is found throughout North Africa, Europe, and across Asia to the Far East. When the time comes for its larvae to pupate, it commandeers orb-weaver spiders to spin a special web for its cocoons. In Japan, the unfortunate victim is a small spider called *Cyclosa argenteoalba*. A normal spider without a wasp overlord produces two types of web. One is the classic spider web with the circular, sticky, spiral threads used for trapping insects. The other is a minimalist web that the spider weaves when it moults its skin. Arachnologists call this the resting web. It has the frame and radiating spokes of the regular web, but not the closely spaced spiral. A moulting spider doesn't want to trap prey. While it waits for its soft body to emerge from the hard, crusty old skin, it is vulnerable and can't do much. The resting web is no more than a spider hammock as it takes time out from regular life.

A female nielseni wasp climbs up the non-sticky anchor lines of the spider's resting web and pretends to be helplessly caught. Once the arachnid unsuspectingly approaches close enough to its 'meal', the wasp wrestles the spider, often knocking it to the ground. It lays an egg, glues it to the arachnid's body, and flies away. The spider's fate is sealed even as it scampers back to its web.

Once the worm-like larva hatches, it drills a hole into the spider's abdomen and feeds off its body fluids. The spider continues its normal life even though it is saddled with a parasite. When the wasp larva is ready to pupate, it manipulates the spider to weave a web to hold its cocoon.

Since cocoon-holding webs are similar to resting webs, do wasp larvae exploit spiders' innate skill for their own ends? Or was the resemblance mere coincidence? Keizo Takasuka from Kobe University, Japan, and his colleagues pondered these questions.

The researchers visited shrines in the cities of Tamba and Sasayama between mid-April and mid-May to collect thirty-six spiders and their webs. By taping each web's anchor lines to a round wooden hoop, they removed the webs from the field. In the lab, they struggled to keep the spiders alive.

'The most onerous obstacles were the rarity of finding wasp larvae on spiders and rearing the spiders,' says Takasuka.

'Many spiders wouldn't weave indoors. They starved and eventually died. The larvae also died. So, I found another method. I kept many unparasitized spiders of same species in the university campus, and then replaced the spiders with precious parasitized ones. They settled down in the already constructed webs. This method provided me with several wasp larvae that reached pupal stage and manipulated the spiders.'

The researchers say the cocoon webs are modified and reinforced resting webs, with structural strands but no sticky spiral. When spiders go into a moult, they produce a specific hormone that guides them through the building of their resting web. Takasuka speculates the larvae must inject a similar hormone-like substance into the spiders, which must instigate the hosts to build much more than a resting web. The immature wasps somehow commandeer the spiders to work an average ten hours at night, going back and forth and laying more silk filaments to bolster the structure.

Although the silk filaments for cocoon webs did not differ from regular webs, the reinforcement made the cocoon webs much tougher. Their frame is almost three times stronger and the spokes forty times tougher than average resting webs. A spider takes only two days to moult, but a pupating wasp requires ten. A cocoon web has to last the duration.

In addition, the parasitized spiders also add finishing touches: little fluffs of silk that flutter in the wind along the spokes of the web. This is a standard ornament in resting webs too. Entomologists cannot agree whether the role of these appendages is to deter predators, attract prey, or advertise the web so birds and large insects don't unwittingly crash into it.

Since the silk fluffs are clearly visible under ultraviolet light, the researchers speculate they may function as 'Do Not Disturb' signs. The last thing moulting spiders and pupating wasps want is a web crasher. The classic trapping web with a spiral has no such decoration and is invisible under ultraviolet light. After all, spiders don't want to alert their insect prey.

'One possible new question for this species, and in fact for all the others that have been studied, is whether the wasp larva may induce the spider to modify the chemistry of the silk lines themselves, to produce stronger fibres,' says William Eberhard of the Smithsonian Tropical Research Institute, Costa Rica. 'This kind of change, in response to environmental factors, is beginning to be documented in other species of spiders, but has never been looked for in the cocoon webs of parasitized spiders.'

Once their jobs are done, the spiders return to the centre of the web. The ill-fated arachnids serve their masters well, feeding them and constructing safe houses. The larvae suck their spiders dry, and then, having no further use for them, chuck the carcasses away. Almost immediately, the wasp larvae build themselves a cocoon by producing silk of their own.

One thing prevents most spiders from suffering this awful fate of becoming zombies – these wasps are relatively rare.

10

Making Peace with Wolves[1]

LARGE troops of adult geladas (*Theropithecus gelada*), a baboon-like primate, dot the high-altitude grasslands of Guassa plateau in central Ethiopia. They graze on grass and herbs and dig up bulbs. During the middle of the day, a lone Ethiopian wolf (*Canis simensis*), no bigger than a medium-sized dog, joins them. As it wanders among them, staring at the ground, an observer would expect the baboons to flee. Instead, they pay no attention even when the reddish-coloured canine drops into stalk mode. Moments later, ignoring gambolling young baboons nearby, the wolf pounces on a rat. Even then, the geladas, the same size as an average wolf, don't flinch.

With a total population of less than 450 animals, the Ethiopian wolf is the most endangered canid in the world. It is usually a lone hunter of rodents. Occasionally, packs of two to thirteen animals hunt larger prey such as antelope and sheep. They could take a gelada, but they don't seem interested. With impressive canines, adult gelada males are no lambs. Still, it is folly to let a predator in their midst.

Although resident Ethiopians may have observed this intriguing behaviour, the first to study it were Vivek Venkataraman from Dartmouth College and his colleagues from other universities and organizations in the US and Ethiopia. For eighteen months, they observed geladas' reactions to wolves and dogs. The primates allowed wolves to approach within four metres. 'Smaller individuals are more likely to move away from a wolf when it gets close,' said Venkataraman. 'The larger animals don't care so much.' Wolves and geladas didn't accidentally hang out together. Their association seemed deliberate and lasted at least an hour at a time.

This reminded me of a similar event at Katagamuwa, adjacent to Yala National Park, Sri Lanka. Large mugger crocodiles swam in tight circles in a shrinking pond. Every now and then, one raised its head above the water to gulp down a big snakehead fish, whose belly flashed gold in the early morning sunlight. Anxious smaller muggers splashed muck as they got out of the way of larger crocs.

A dozen painted storks strutted amongst these toothy, armour-plated predators, hunting for fish of a size they could handle. Unlike smaller crocs, the storks didn't show the slightest nervousness of the large predators. If the foolhardiness of the birds amazed me, I was even more astonished by the crocs. I watched them hour after hour for days, and they seemed to have an eye only for fish. Katagamuwa is exceptional for the number of congregated crocs, at least 150 large ones.

Just as the crocs ignore the spindly pink legs stepping tantalizingly near their faces, the Ethiopian wolves give up larger prey in favour of tidbits.

However, when domestic dogs from nearby settlements trotted up, the monkeys shrieked and grabbing their babies fled up cliff slopes. Despite the geladas' dramatic precaution, observers watched dog packs kill three monkeys on different occasions. But, the researchers saw only one wolf attempt to catch a baby gelada. Dozens of bold monkeys mobbed the wolf until it dropped the baby unscathed.

Obviously, the primates didn't wait until dogs got close before scurrying away. The authors say geladas distinguished wolves from dogs even though they were of similar size and colour from at least 300 metres away. I suspect it is also a matter of numbers. A lone wolf is not as threatening as a pack of dogs. Should a pack of wolves join the geladas, no doubt they would flee.

Why do wolves prefer to fraternize with geladas?

In the middle of the day, when rodents see a bunch of geladas stuff themselves with grasses and bulbs, they may assume there is no danger and let down their guard. Or perhaps, the authors speculate, geladas disturb rodents while foraging and flush them out. Or maybe rodents can't hear or see anything else above the noise of geladas crunching and munching. Whatever be the reason, wolves exploit the cover provided by the primates to feast on their favourite prey. Researchers say that without a gelada escort, wolves are less successful.

With much to gain, wolves seem to do their utmost to win the geladas' trust. Not only do they forgo many opportunities of taking a gelada baby or two, they even behave cautiously near the monkeys. They stalk slowly when the primates are nearby. When geladas aren't around, the predators would trot in a zigzag fashion across the grassland. They also hunt alone to avoid spooking the primates, say the authors. But wolves don't need to form packs to bring down small rodents.

What do geladas gain by letting wolves into their midst? Do rodents compete with geladas for the same food? Are geladas using wolves to get rid of the competition?

'Rodents are in a sense a competitor [to geladas],' says Venkataraman, 'because they consume similar resources, but this is difficult to test. I think geladas tolerate the wolves because the cost of repelling them constantly is too high. They're not generally threatening, so they let them be around. As far as we can tell, the geladas derive no benefits from the presence of the wolves.'

If geladas don't gain anything from wolves, what about the storks who hunt with crocs?

I surmised the predators have an unspoken treaty to amicably split the seasonal spoils. Storks' feet stir large fish out of the mucky bottom into croc jaws, and in return, crocs let the birds go unmolested. Since they hunt fish of different sizes, they aren't competing with each other. Still, ignoring a large stork or gelada meal for little morsels displays self-restraint and an ability to make trade-offs, not qualities one associates with crocs and wolves, the bad boys of many of our tales.

The losers in these unlikely pacts, however, are the small fry: fish and rodents.

11
The Amorous Octopus[1]

THE antics of octopuses might be amusing, but they are extremely nasty to each other. Even when mating, they don't canoodle. That is because these solitary creatures are cannibals. The large females can easily overpower and eat their smaller suitors. Some males avoid becoming a meal by leaping on the females' backs and out of their reach. Others keep their mates at arm's length and shove little packets of sperm into oviducts located inside females' mantles, the round bell-like octopus body. Should females show any sign of aggression or hunger, males flee. One species is so petrified of being eaten, the males detach their sperm-laden arms and offer them to females. But there is one exception.

In 1975, Panamanian marine biologist Arcadio Rodaniche was surprised that larger Pacific striped octopus didn't copulate like other octopuses but more like squids and cuttlefish. He watched in amazement as mating pairs snuggled up together in the same den for a few days like honeymooners and had intimate sex by locking mouths and linking their sucker-lined arms. They even shared food mouth-to-mouth like lovey-dovey humans, a suicidal move for males of other octopus species. The pale-coloured female enveloped her darker mate, and he responded by jetting ink. No other octopus behaved so sensuously. While female octopuses of other species died soon after laying eggs, larger Pacific striped octopus mothers survived for months afterwards.

'I started working with them by accident,' says Rodaniche. 'In the late 1970s, a friend kept picking up the larger Pacific striped octopus in his shrimp trawler. He asked me if I was interested. He was making short trawls, so the animals arrived in very good health.' Rodaniche studied the species in a large saltwater swimming pool at the Smithsonian Tropical Research Institute in Panama.

However, the scientific world turned its back on his work. The behaviour of the larger Pacific striped octopus upended everything known about octopuses. 'I presented the original paper on the larger Pacific striped octopus at a symposium at Woods Hole in 1990,' says Rodaniche. 'Although some found the data intriguing, it was clear that there were others who flatly refused to accept it. In December that year, I learned that the paper had been given a very unfavourable review by the only person who reviewed it. They [the editors] gave me the opportunity to revise it, but because they required me to make many substantive changes in a very short period of time, I declined. I retired from the Smithsonian three years later.' No one paid any further attention to these animals that go against the grain of octopus behaviour.

In 2012, Richard Ross of California Academy of Sciences and Roy Caldwell of the University of California, Berkeley, US, bought twenty-four specimens from a pet dealer. He had collected the animals in sandy soil at a river mouth along the Pacific coast of Nicaragua, where they live in colonies of up to forty at depths of forty to fifty metres. When Caldwell heard from the collector that all the animals came from a congregation of octopuses, he suspected Rodaniche was right.

The biologists observed the octopuses at the Steinhart Aquarium at the California Academy and at the university's Berkeley campus lab.

Despite its name, the larger Pacific striped octopus is not big. Females are the size of plums, while males are lime-size. They grab crabs, pry mussels and clams apart, and drill holes into snail shells to extract the soft flesh, like any other octopus. But they add a bit of drama when hunting shrimp.

When a larger Pacific striped octopus spots a shrimp in the distance, it creeps up behind it, arches an arm over it slowly, and taps it on the far side. The startled shrimp jerks backwards into the waiting arms of the octopus. Caldwell says the reason for this unusual hunting strategy could be shrimps' ability to swim fast. The octopus cannot hope to catch its prey by chasing after it and hence resorts to subterfuge.

In the labs, the cephalopods mated sensuously just as Rodaniche had described. Only once did a female turn aggressive, when a male attempted to mate with her. She may have been brooding, but other brooding females offered no resistance to amorous males.

Female octopuses have two oviducts, so they can mate with two males simultaneously. The scientists speculate that larger Pacific striped octopus may have evolved this unique manner of beak-to-beak mating so that the male can monopolize the female and prevent her from mating with another.

The females lay an egg a day for up to six months and stick them in pairs to walls of the den. Throughout incubation, they tend the eggs by running their arm tips through them.

When researchers entered the room, wandering females immediately returned to their cache of eggs until the humans left.

Three months later, hatching begins. While other octopuses starve to death after producing a single brood, female striped octopuses continue to eat, mate and lay hundreds of eggs.

'Only by observing the context in which these behaviours occur in the wild can we begin to piece together how this octopus has evolved behaviours so radically different from what occurs in most other species of octopus,' Caldwell said.

The octopuses in California died naturally after reproducing. After they abruptly stopped eating, males died within two weeks. Once females reached the end of their lives, they ate a little bit, less frequently, and gradually retreated to a corner of the aquarium. If they had eggs stuck to the wall of their dens, they continued to care for them. They died between two and four months later.

Caldwell says the biggest problem the research team faced in studying the species was getting specimens. 'They are extremely difficult to collect due to depth and visibility.' He hopes to continue studying the species by filming it in the wild.

'I feel vindicated knowing that my data was finally corroborated by other scientists,' says Rodaniche. 'Because the octopuses' behaviours were so unusual for what was then known about them, and because the animals are not easy to locate, I decided that it would only be a matter of time until someone else found them and recognized how unique they are.'

Despite its unusual behaviour, the octopus has not been scientifically described and named. We know this unique creature only by its insipid name: larger Pacific striped octopus.

12

Pregnant Fathers[1]

SEAHORSES, pipefishes, and seadragons are the only males in the animal kingdom to get pregnant. Do their brood pouches serve merely as safe, enclosed nurseries for their young? Or do they provide nourishment? What happens inside the pouch was a mystery until recently.

Male seahorses have pouches in their abdomens. In the seas of southern Australia, the foot-long male big-belly seahorse (*Hippocampus abdominalis*) courts his lady by making his belly look even more rotund. He fills it with water and accentuates its colour by turning it yellowish-white or white. If she is enamoured, she reciprocates by contrasting her body colours. Facing each other, they swim together, sometimes with their tails entwined, looking like animated dancing balloons.

When the time comes for mating, the male urges the female to float up to the surface with him by pointing his snout upwards. She signals her readiness by tilting her head up, and they drift to the ocean surface in tandem. The female presses her abdomen against his and spews eggs into his open pouch. The male releases sperm to fertilize the eggs as they enter.

Camilla Whittington of the University of Sydney, Australia, and her colleagues took tissue samples from male big-belly seahorses throughout their pregnancies.

'These seahorses have a big belly whether they are pregnant or not,' says Whittington. 'It took us quite a while to figure out how to tell if they were pregnant. It's not as easy as in humans, who can just urinate on a stick. We ended up developing a method of pregnancy testing based on their behaviours. We realized that pregnant dads did not do the normal courtship dance with females that non-pregnant males did.'

To conduct this kind of study, the researchers had to know much more than whether a male seahorse was carrying eggs. They had to figure out the stage of pregnancy.

Soon after mating, the male seals the opening of his pouch so seawater cannot enter. Extremely protein-rich fluid floods the pouch, and its outer layer, the epidermis, becomes thick with blood vessels. These developments are essential for gas exchange, and to excrete the nitrogen the embryos generate. Human placentas do the same job.

Embryos live on egg yolks provided by the mother. Through the 24-day gestation, their father provides additional nutrients, such as energy-rich fatty acids called lipids, calcium for their skeletons, and iron to absorb oxygen. They may be produced in the pouch itself.

All these nutrients swishing in an enclosed space are also an excellent medium for bacterial growth. The male seahorse protects his embryos by secreting anti-bacterial proteins and antifungal peptides.

The father himself could be a threat to the embryos attached to his body, if his immune system recognized them as foreign bodies and rejected them. Therefore, genes that cause inflammation are suppressed so they don't trigger his immune system.

Toward the later part of pregnancy, the pouch gradually absorbs seawater so the babies become used to salinity. This flushing brings in more oxygen for the growing young and removes their excrement.

About a week before hatching, the father sends out 'hatch now' signals. The embryos emerge from the thin membranes enveloping them and swim around within the pouch. They continue to live off the orange yolk sacs attached to their bellies. As they grow, the babies stretch the pouch to the limit.

Although the brood pouch flap is no more than modified stomach skin, it's not all that different from a placental mammalian uterus.

On a full moon night, the father produces isotocin, the fish equivalent of oxytocin, which induces labour. Rhythmic muscular contractions push the babies out. The young seahorses, measuring the length of a clove, drift off into the sea. The bigger the father, the more babies he has. An average-sized dad may produce 300 to 700 young at a time. The most any big-belly seahorse dad has produced is 1,100 babies.

Millions of years separate seahorses from mammals, reptiles, and even live-bearing fish. Despite the aeons of separation and independent evolution, a pregnant seahorse father undergoes similar bodily changes as a pregnant woman. That's because the needs of the young are the same: oxygen and nutrients.

After the last of the young leave, the pouch changes structure rapidly, and the father becomes ready for the next round of courting and bearing young. He'll get pregnant three more times before summer is over.

Father seahorses enjoy one reward for taking on the young-bearing burden: they have no doubt of the paternity of their offspring.

13

Why Don't Elephants Get Cancer?[1]

ELEPHANTS, African and Asian, rarely fall victim to cancer. But humans aren't so fortunate. Close to eight million people die of cancer annually, including 1,500 Indians every day.[2] While about 95 per cent of cancer can be prevented by improving our lifestyles,[3] a good many people suffer from it because of genetic defects. Understanding how elephants beat the dreaded disease can hold clues for treating human patients. Researchers have taken the first step in that direction.

In less than ten years, 100-kilogram newborn elephants put on more than 3,000 kilograms in weight, a thirtyfold increase in cell mass. Such a high rate of cell multiplication in a large animal is prone to DNA copying errors. Since elephants are long-lived, up to seventy years, these mutations should accumulate and cause malignant tumours. With such risk factors as body size and age working against them, the world's largest land animals ought to be dying of cancer all the time.

In the 1970s, Richard Peto, at the University of Oxford, observed that only a few of these giants appear to develop cancer. However, animals at the other end of the body size and age spectrum such as mice and rats are highly prone to the disease. This contradiction between size and susceptibility to cancer has become known as Peto's Paradox.[4]

In 2012, Carlo Maley, an evolutionary oncologist from Arizona State University, gave a talk about Peto's Paradox at a conference of evolutionary medicine and epidemiology in Washington, DC. He had found that elephants had twenty copies of a gene called TP53 that produces p53, a protein crucial for suppressing tumours.

This protein is like a fact-checker, examining newly formed cells and verifying if they have correct DNA copies. If any are corrupted, it either attempts to repair or gives the command for the cells to commit suicide. If it didn't issue these instructions, the cells could turn cancerous. The protein is so crucial to our well-being that David Lane, a British cancer biologist and one of the discoverers of p53, called it the 'guardian of the genome'.[5]

Humans have one copy of TP53. Every gene has two alleles. Both of them have to be in working order to prevent cancer. When people inherit only one functional allele of TP53, they have Li-Fraumeni Syndrome (LFS), which leaves them with nearly 100 per cent lifetime risk for the disease.

Do multiple copies of the DNA fact-checking gene shield elephants from cancer?

Joshua Schiffman, a paediatric oncologist at Huntsman Cancer Institute and University of Utah School of Medicine, Salt Lake City, was in the audience at Maley's talk. Schiffman treats children genetically predisposed to cancer and was especially interested in the elephants' adaptation. He collaborated with Maley and eleven other colleagues to investigate the role of p53 in cancer prevention.

To verify the incidence of cancer among the pachyderms, the team turned to *The Elephant Encyclopedia*, a database of almost 6,000 captive elephants in 2,000 locations across the world. By analyzing the cause of death of nearly 650 animals, Schiffman and his colleagues concluded that 5 per cent of captive elephants died of cancer compared to 20 per cent of humans. So Peto's observation checked out.

The researchers then took white blood cells from eight zoo elephants, eleven normal humans with no family history of cancer, and ten patients suffering from LFS. They tested how the cells responded when their DNA was damaged.

Cells of people with LFS were not as sensitive to damage as normal human cells. In fact, they were less than half as responsive. Their p53 was too weak to order the destruction of cells, which, therefore, had a higher chance of turning cancerous.

Elephant cells were twice as responsive as normal human cells. Elephantine p53 didn't stop to repair the damage. Instead, it sent out instructions to destroy the corrupted cells, as if it didn't want to take any chances.

In another study, researchers, led by evolutionary biologist Vincent Lynch of the University of Chicago, traced the evolution of TP53 using DNA from fossils of mastodons and woolly mammoths.[6] Pachyderm ancestors had small bodies and had only one copy, but as their descendants grew in size, so too did the number of copies.

Lynch and his collaborators subsequently discovered another cancer-fighting mechanism in elephants.[7] The pachyderms have ten copies of a gene called Leukemia Inhibitory Factor (LIF), while other mammals have only one. As its name indicates, LIF suppresses the growth of blood cancer cells and also promotes bone and nerve formation. It plays a role in implanting the embryo on the uterine walls of human mothers.[8]

In an ancestor of elephants and their relatives, like manatees and hyraxes, the duplication process went crazy, creating multiple copies. That doesn't by itself mean that these animals have that much more power to suppress tumours. As often happens, such duplicates sit in the DNA like duds, being faithfully copied from one generation to the next and doing nothing for the functioning of the organism. That is because regulators that activate these copies are missing. As evolution began bulking the bodies of proto-elephants about thirty million years ago, it also developed a regulator to resuscitate one copy. This reanimated gene became what researchers call LIF6.

The key that turns on LIF6's regulator is p53. When p53 comes across an improperly copied cell, it triggers LIF6 to release proteins that bore holes into the mitochondria of the target cell. Smaller proteins called cytochrome c pour out of the mitochondria, setting off a cascading series of reactions that kill the cell.

Now that researchers know some secrets of how elephants resist cancer, what are the next steps?

Having multiple copies of LIF6, however, is not an option since they would kill healthy cells. Schiffman's team plans to explore two approaches. 'One is finding a drug, either natural or created in the laboratory, that can mimic the same effect of elephant TP53. And the other is working with new genomic technology to specifically deliver elephant TP53 where it needs to go in a patient with early or advanced cancer. We will need to understand both the benefits and risks of using elephant TP53 as therapy for humans, but we're very excited to move forward with our research to learn how it might be able to benefit our human patients.'

Having multiple copies of p53, like LIF6, has its risks. When laboratory mice were given more p53, they became extraordinarily resistant to cancer as expected, but aged and died prematurely. Perhaps, elephants counteract the side effect with another gene that gives them long life.

But p53 isn't the only ticket to a cancer-free life. Although rodents are susceptible to cancer, blind mole rats and naked mole rats are exceptions. Scientists haven't seen a single incidence of cancer in necropsies of 250 naked mole rats.[9] But these rodents have evolved other proteins called p16 and p27 that appear to play a more prominent role than p53[10] in beating the disease. Naked mole rats also produce a substance called hyaluronan between their tissues that shields them from cancer.[11]

Scientists heralded these discoveries as the future of cancer therapy. Should research focus more on p53 than these other proteins?

'Extra p53 seems to be unique to elephants,' says Schiffman, 'but there almost certainly are different genes and pathways that other animals use to decrease their own risk of cancer. P53 is involved in over half of all human tumours, and loss of functioning p53 leads to Li-Fraumeni Syndrome. We are focusing on p53 because of its importance in cancer regulation in humans, and now what we believe may be its potential role in cancer prevention in elephants.'

Even as scientists hunt for a way of using p53 to beat cancer, they may also need to find the associated gene: the one for long life.

14

Sticklebacks Hold Their Water[1]

MALE stickleback fish seem to hold their urine for the length of their breeding season. And yet, they don't seem to be in pain or suffer symptoms of waste buildup. They continue fighting with other males and soliciting females with no discomfort. What would be a medical emergency in humans is just a phase these fish go through.

Sexually mature male three-spined sticklebacks (*Gasterosteus aculeatus*) stop urinating as soon as the breeding season starts in late spring and early summer. Until then, they pee like other creatures. Male sex hormones are to blame for this dramatic change. Instead of enlarging the testes, these hormones cause the kidneys to become visibly bigger.

The primary role of kidneys is to filter blood and excrete waste as urine. If they don't produce urine, what do this fish's large kidneys do? From excreting organs, they are reassigned to become secreting glands.

For the five-centimetre-long male sticklebacks, nests are the ticket to sex and progeny. Females assess how good a mate a male is by the nest he makes. Three-spined sticklebacks live in the seas, lagoons, and freshwaters of the northern hemisphere, and they adapt their nest-building technique according to the situation. Typically, hormone-charged males construct a depression in a sandy streambed, arrange plants and filaments of algae on it, and then create a tunnel through the vegetation. They cement the mass of plants with strands of a viscous glue-like protein called spiggin, produced by the fish's kidneys.

With such a critical job to do, the kidneys forsake producing urine altogether and instead secrete the slimy spiggin that is then stored in the urinary bladder.

The problem becomes acute for sticklebacks living in freshwater. Since the fish's blood has much higher salt content than the surrounding water, they absorb water. The primary function of the kidneys is to get rid of this surplus water. During breeding season, when the role of the organs changes entirely, the fish has to void it somehow.

'If he cannot get rid of the excess water, he ought to swell up like a sponge, absorbing water, and eventually explode,' says Steffen Madsen of the University of Southern Denmark.

How do male sticklebacks rework their internal plumbing to get around this problem?

When the kidneys switch to spiggin production, the male stickleback's intestines also change. They take over the job of excreting fluid, but scientists didn't know how until now.

'Our research now gives a good and detailed clue to how this is possible,' explains Madsen.

Madsen and his team studied intestinal cells of the three-spined stickleback in the lab and analysed several proteins involved in transporting salt and water. They discovered how the intestinal wall excretes urine through specific molecular channels called aquaporins. From there, the waste liquid is directed to the cloaca, the fish's excretory and reproductive orifice. Taking on the additional duty of eliminating water subtly changes the structure of the intestinal cells.

The fluid that intestines secrete is plain water. For fish living in freshwater, salt is a precious commodity that they cannot afford to waste. So, the intestines do not process salt and minerals.

'Other waste products, like ammonia, are easily excreted by the gills,' says Madsen.

Male hormones enlarge the kidneys of male and immature female sticklebacks. For nearly a decade, scientists have suggested this amazing physiological change in these fish could be used to detect doping by athletes or testing water quality for effluent and agricultural runoff.[2] Steroid-tainted liquid makes the fish grow enlarged kidneys. Some pesticides and fungicides prevent male sticklebacks from producing spiggin. The fish could be aquatic environmental sentinels, like canaries in a coal mine. Since these chemicals disrupt hormonal activity in other aquatic creatures, they could alert us when pollutants reach unacceptable levels in streams, rivers and seas.

Spiggin production is seasonal. Once the nest is made, the male courts the female by performing a zigzagging dance. If she shows interest, he points his nose to the entrance of his tunnel nest. After she enters and lays her eggs, the male follows and fertilizes them. Then he courts other females successively to lay their eggs in his nest, sometimes even building multi-tiered nests to accommodate them.

Once he's done with nest building, he tends the eggs, protecting them from other fish. Since the eggs suck oxygen from water, he continually aerates them with his fins. When the brood hatches, he defends the babies until they can hunt worms and insects on their own. During this time, the male's sex hormones slowly decline. Before his kidneys can return to shape and resume the job of urinating, the male plunges into the next breeding cycle. He goes through three bouts of reproduction in two months before summer ends and his kidneys finally return to normal functioning.

No other fish is known to excrete through the intestines.

15

'Who am I?'[1]

SOME birds outsource parental responsibilities. Investing tremendous effort in raising their chicks is not for them. Brown-headed cowbirds (*Molothrus ater*) of North America surreptitiously lay their eggs in the nests of other species like prothonotary warblers (*Protonotaria citrea*). The foster parents raise these interlopers along with their own, freeing the parasitic biological parents to live their own lives. Despite the clear advantages, only 1 per cent of all birds force other species to provide free parental care called brood parasitism.

Chicks raised by parents of another species could face an identity crisis. When their foster parents and siblings belong to a different species and don't look or act like them, how do the free-loading chicks know who they are?

Chicks learn songs, diet, social etiquette and migratory routes from their parents. Adopted chicks can't learn these from their foster parents. Do the youngsters pay the price for their biological parents' carefree life?

The brown-headed cowbird is a fruit and insect-eater, but warblers eat insects and caterpillars. Adult cowbirds live in open fields and prairies while warblers live in forests. Do cowbirds warble their own song or do they sing like warblers?

Adult cowbirds seek drab cowbird mates and not bright yellow warblers, illustrating they know who they are. Could cowbird mothers play a role in grooming their offspring? This is not a far-fetched idea. Cowbird mothers don't just drop their eggs in warbler nests and abandon them to their fate. They keep an eye on those nests. If the surrogate parents shove the strange eggs overboard, cowbird mothers vandalize the nests, forcing the hosts to build another one.[2] This threat of violent retaliation leaves warbler parents no choice but to incubate cowbird eggs. When their young beg for food, cowbird mothers call in response. By looking at the DNA, the authors of one study found 36 per cent of adult and young cowbirds caught in the same trap were closely related.[3]

How do these young cowbirds find their kin or flocks of their kind?

Biologists from the University of Illinois at Urbana-Champaign and University of New York, US, found the answers to a small part of the puzzle.

At first, Matthew Louder, the lead author, and his team speculated that cowbird mothers, hanging around the nests, may come to the aid of their offspring during fledging time. Even if parents didn't actively encourage their offspring to follow them, perhaps the youngsters became fixated on their real mothers, whom they may have seen around the nest, and followed them.

Mark Hauber, one of Louder's fellow researchers, suggested a password specific to the species could be the key.[4] He figured a specific type of chatter made by female cowbirds is the secret code that yanks their chicks to attention. When he played a recording to six-day-old nestlings, they begged more eagerly than to any other birdcall. Two-month-old fledglings didn't hesitate to approach the device playing the chatter, suggesting that the password enables the offspring to recognize their real mothers. Hearing the coded call for the first time triggers learning. The youngsters could pick up cowbird behaviour from the birds making the chatter.

To test how this works in real life, Louder and his colleagues fitted radio transmitters to fifteen fledgling cowbirds and fifteen adult female cowbirds found near the nests.

Louder followed one individual at a time from morning until evening. If it flew, he ran after it through the forest, sometimes for as much as a kilometre, returning home exhausted every evening. 'Like a naïve parent, I assumed that these teenagers stayed put all night,' says Louder.

Michael Ward, one of Louder's collaborators, came up with another approach. He constructed an automated system of three radio towers, each with six antennas, which provide 360° coverage. The system tracked one bird at a time, every one to two minutes.

To the researchers' surprise, the youngsters flew out alone every night from their nests in the forest and roosted in the open fields, without any adult supervision. Although some female cowbirds hung around in the woods at fledging time, they weren't around when the young wanted to leave. Neither were the offspring within the home range of their mothers.

Cowbird mothers didn't play a role in helping their young fledge.

When Louder analysed the data, he thought something was wrong with it. Perhaps the chicks were roosting on the ground instead of up in the tree, and that was garbling the automated telemetry system. By this time, only two remaining youngsters hadn't become independent. He raced after each bird for several nights, as it flew out of the forest alone to the fields. The following morning, the fledglings were back at their nests, begging for food from their warbler parents. The automated system had been correct.

The researchers speculate the juveniles may have an innate preference for specific roosting habitats. Perching alone at night may help young cowbirds become independent of their foster parents. But unaccompanied underage birds were taking too many chances with predators.

'This is what makes this behaviour so peculiar,' says Louder. Not only do the birds venture out alone at a dangerous time, but they frequent places they have never been before. 'This is one reason that led me to believe this behaviour is important. Otherwise, it would not persist.'

When cowbirds spend more than a month in the nest, they think of themselves as warblers. To avoid imprinting, they distance themselves from their surrogate parents by heading for the fields every night. But this behaviour also provides opportunities to find flocks of the same feather. Adult cowbirds fly to communal roosts in groups every evening. When the fledglings sit night after night alone in the fields, they listen for this password chatter to find such flocks.

In a subsequent study,[5] Louder and Hauber, collaborating with Kathleen Lynch, an avian neuroscientist, analysed calls and examined the brains of juvenile and adult cowbirds.

Unlike warblers that learn to sing as chicks, cowbirds are late learners. They pick up songs only when they are two years old and sing with expertise at three. The delayed development ensures they don't learn warbler songs.

While coercing others to take care of one's young appears to be a lazy way of parenting, the chicks have to work harder. They must brave predators and fly back and forth to avoid becoming fixated on their adopted parents. At least they can look forward to adulthood with few childcare responsibilities.

16

Rodent Monogamy[1]

PRAIRIE voles (*Microtus ochrogaster*) are said to be the paragons of monogamy. These brown mouse-like creatures, smaller than a palm squirrel, are among only 3 to 5 per cent of about 5,000 mammals to stick to one partner all their lives. When one of a couple dies, its mate most often lives a solitary life. But some studs give the lie to their species' monogamous lifestyles. They trespass upon other voles' home ranges, sneaking sex and scrapping with strangers. Female partners in stable relationships aren't above having a casual fling. Their regular mates don't father even as much as a quarter of the litters. Still, a majority stick with their first sex partner, and that's a rarity in the animal world.

Philandering males of these small North American rodents hit the reproductive jackpot, producing more babies than their monogamous counterparts. But being unfaithful is not all fun and frolic. While these males are gallivanting, their lonely mates may also seek sex with others. These cuckolded males that are busy cuckolding others are more likely to raise a rival's offspring. Neither monogamy nor polygamy has an edge, and prairie voles practice both.

What makes some stick to one mate and the others wander?

Vasopressin, a hormone produced by the brain's hypothalamus, influences social behaviour such as aggression, bonding with mates and care of offspring. It suffuses the brains of monogamous voles that have more vasopressin receptors (V1aR)[2] in the ventral pallidum, the reward-seeking part of the brain.[3] It is the same with faithful men.[4] When receptor-rich voles have sex for the first time, they imprint on their partners and thereafter stay true to them.[5]

Meadow voles (*Microtus pennsylvanicus*), closely related to prairie voles, are promiscuous. By injecting vasopressin receptors into the males' brains, researchers turned them monogamous.[6]

The receptors were not the only things keeping male prairie voles in line. A lot has to do with their sense of geography, say Steven Phelps of the University of Texas at Austin, US, and his colleagues. They discovered rodents with more vasopressin receptors in the brain's retrosplenial cortex, known for its role in memory and navigation,[7] stayed within the confines of their homes.[8] These rodents recalled sites where they had fights with rivals and avoided them.

The ones with fewer vasopressin receptors had a poorer understanding of their territories. They wandered over large areas, intruding on others' territories and perhaps only dimly recollecting where their nests were located. The result: promiscuity.

Why didn't the prairie voles stick to either monogamy or polygamy? Mariam Okhovat and other students in Phelps' lab used genetics in the lab combined with radio telemetry data gathered in a previous study to find out. They examined the gene avpr1a, known for predicting sexual fidelity in males and encoding the vasopressin receptors V1aR.

The specific arrangement of the four nucleotides in a DNA sequence is called encoding. It is the recipe for producing a particular protein that plays a unique role. In this case, the gene avpr1a produces a protein that acts as a receptor for vasopressin.

Not only did the number of V1aR in the ventral palladium and retrosplenial cortex anticipate male behaviour, but variations in the sequence of the gene also mattered.

The researchers discovered avpr1a comes with two alleles that they call 'HI' and 'LO.' The structure of the V1aR produced by the HI and LO alleles is the same, but they produce different amounts, says Okhovat. Males with the HI allele producing more vasopressin in the retrosplenial cortex were faithful and stayed within the bounds of their home range. The ones with the LO allele produced low levels of the hormone in the retrosplenial cortex and bumbled far out of their territories. Although the researchers noticed this correlation, this relationship between geography and vasopressin in the retrosplenial cortex hasn't been confirmed.

The two alleles didn't affect the amount of V1aR receptors in the ventral palladium, which regulates motivations, emotions and behaviour. The gene variations only affected receptors in parts of the brain involved in remembering places and sexual fidelity.

Phelps says his team was shocked by what they found. 'This brain variation isn't just there by chance. It isn't random.'

Natural selection had maintained the alleles in prairie voles for a long time so they had to benefit the species.

Since these effects are stronger in captive voles than in wild ones, the researchers say the traits cannot be explained by heredity alone. Environment and population may also influence the degree of males' faithfulness.

Like many rodents, their populations boom or crash depending on food supply. Numbers explode at feast times, and home ranges shrink in size, with more than 600 rodents jammed in a hectare. Such situations are a perfect playground for promiscuous voles with little sense of territory, and a headache for possessive male voles, keen on keeping their mates away from strangers' charms.

During famines, vole populations crash, leaving only one or two in a hectare. Those with better navigation and memory skills are at an advantage. They dote on their partners, defend their territories and mates from philandering males, and make successful parents.

Extending these conclusions to human behaviour is tempting, but research hasn't advanced that far. A 2014 study[9] of 7,400 twins in Finland is the latest to suggest genes could influence faithfulness. It is possible vasopressin receptors control fidelity in women and, to a lesser degree, in men. Surprisingly, oxytocin, commonly thought to be at work when women bond with their partners, seems to have little effect. Philandering in humans may have some biological basis, but it is complicated by personality and environment.

Of his findings in voles, Phelps says, 'When it comes to social behaviour, maybe there isn't a normal brain.' Many genes may have subtle variations so that individuals of a species can respond with flexibility to circumstances.

Wandering voles go astray in more ways than one.

17

The Snake that Dodges Predators?[1]

Puff adders (*Bitis arietans*) lie in ambush throughout sub-Saharan Africa for toads, rodents and lizards. They may be dreaded, venomous assailants of these small creatures, but the fat, sluggish metre-long vipers are delectable sausages for larger animals. Researchers uncovered the species' extraordinary trick to avoid being eaten.

Unlike the sit-and-wait adders, cobras use up energy, going from burrow to burrow hunting rodents. Should they spy a wild pig that could gobble them up, they escape by zipping down a rat hole. The puff adders aren't quick on their feetless bodies and rely on the disruptive dark bands and rough brown scales to blend with the surroundings. None of the vipers' enemies, from cobras to warthogs and raptors, can spot them hiding in plain view. Not moving at all brings them supper and saves their lives.

The puff adders' superb cryptic colours escape researchers' eyes too, making them hard to study. Even ones implanted with radio transmitters that beep their locations are hard to spot. 'When you get close to the snake, the signal loses its directionality,' says Graham Alexander of the University of the Witwatersrand, South Africa. 'I'd often struggle to locate the puff adder, even when I had it down to a square metre or so. I regularly found puff adders under my boots when I was tracking them, and they would do nothing.'

A creature close to the ground must feel threatened by a human stumbling on them. But the vipers continue with their policy of 'do nothing'.

Blending in may fool visual hunters but not those with sharp noses. Dogs, with their incredible sense of smell, help us track criminals, rescue buried people, find bombs and drugs and detect diseases in humans. They aid wildlife studies by finding scats of specific species, even the poop of killer whales out at sea. If dogs can sniff out puff adders so could the snake's natural predators. Snakes may have the advantage because they are not as smelly as mammals unless they defecate or musk in distress.

A few of Alexander's snakes with radio transmitters took up residence in the game reserve ranger's gardens, coiling up in flower beds. The ranger's pet dogs stepped right on the venomous snakes as they rushed to greet him. Even two tame banded mongooses (*Mungos mungo*), famed for their enmity with snakes, didn't notice one of these vipers. 'If puff adders freeze when under threat and many of their predators use scent as their main sense, any smelly puff adder wouldn't last long,' says Alexander. Perhaps, the vipers don't have body odour at all.

Anecdotes such as these aren't enough to prove puff adders don't have smell. Ashadee Kay Miller, a post-graduate student at the university and the lead author, investigated how the vipers cope with snake-eaters with exceptional olfactory prowess. She enlisted Shannon McKay, a professional canine trainer, to teach four dogs, while Miller trained five meerkats (*Suricata suricatta*), a relative of mongooses.

Trainers train sniffer dogs on a specific drug or explosive and to show their handlers in some way whenever they encounter it. Instead of using this method, McKay and Miller taught the animals to scent-match, a skill much harder for the mammals to master. From a scent wheel with many smells, the animals pick the same odour as a specific sample presented by their trainers. 'Effectively, this is what a bloodhound is doing when it tracks someone's trail,' says Alexander. 'You present the dog with an individual's scent and the bloodhound then scent-matches it with the trail.'

For the experiment, the researchers left cotton cloths on top of snakes of six different species for forty minutes. Besides the sit-and-wait puff adders, they used species such as corn snake, house snake and even a night adder, all of which were active, going after their prey or slithering away from predators. To make sure nothing else could influence the results, they included fabric scented with vegetation and cloth left in a clean plastic container with no snakes. The researchers also added a piece of freshly laundered textile to the mix.

The trained dogs and meerkats had no problem scent-matching active hunters, but they failed completely with the vipers that seemed to be odourless. The canines performed better when the researchers presented the smell of a newly shed, funky-smelling puff adder skin.

Since arboreal snakes are active, they can afford to have a whiff of odour cling to them. They poop anywhere and at any time they feel the urge. Puff adders, however, have to be careful of making a stink. Their shed skins reek of lubricating fluid making humans hold their noses. It is no surprise the dogs detected the smelly skin. The snakes defecate soon after moulting and move away to another site. Perhaps, they don't want to hang around where their stinky droppings and sloughs would advertise their presence.

Not all snakes that are ambush hunters evolved this useful trick of being odourless. Growing to over two metres long, eastern diamondback rattlesnakes have far fewer predators. Under little pressure to develop new skills, they aren't shy about smelling funky.

Despite puff adders' superb sight and smell camouflage, Alexander says almost 50 per cent of his subjects die every year. Why does that happen when predators can neither see nor smell them?

Puff adders experience the stirrings of lust during the southern hemisphere's winter, from May to June. Males abandon their sit-and-wait approach and wander yonder range in search of mates. 'They move; they die,' says Alexander. Their cryptic colours provide cover as long as they are still. The urge to procreate takes its toll on females too. To attract mates, they crawl around, marking their trails with musk from specialized glands in their tails. They don't slither far and wide like the males, so most survive.

Based on his fieldwork, Alexander says snouted cobras kill more puff adders than any other predator. 'This might explain why puff adders spend about 95 per cent of their time on the surface under grass. Cobras can easily follow puff adders down holes, and holes would be focal points for cobras to search. But if you stay under an arbitrary tuft of grass, you make things more difficult for the cobra.'

Prey animals stay one step ahead of their predators. Predators try to narrow the gap and see through their preys' tricks. In this evolving competition, puff adders have the advantage albeit one that entails being stepped on by dogs, mongooses and researchers.

18
Food or Mates?[1]

GREAT tits, sparrow-sized European birds, spend most of their waking hours looking for food. Filling up their stomachs takes on more urgency in winter, in the weeks preceding nesting. They lay up to twelve eggs, keep them warm and then cater to the demands of growing chicks until they fledge. Single parenthood is not a choice. Both parents have to be in fine form to withstand these rigours, and their strong bond sees them through the tough time.

Scientists from the University of Oxford and the University of Konstanz, Germany, put the yellow-breasted birds to the test to see which need outweighed the other: food or mate.

Great tits (*Parus major*) feed at feeders set up throughout the Wytham Woods, west of Oxford, UK. The researchers fitted radio-identification transponders, also known as PIT tags, to the ankles of seventeen pairs. They connected the readers to the doors of the feeding stations. Whenever the tits arrive to eat, the device recognizes and records each bird's unique number.

'They are mini, lightweight, cheap and power-free tags that can be detected as the individual animal comes within range of a reading station,' says Joshua Firth, who led the study. 'This allows us to track great tits through their whole life.'

The readers did more than scan identity codes. The researchers programmed half of them to open the feeders for birds wearing odd-numbered tags and the other half for even-numbered tags.

Some couples had it easy. If both birds wore odd or even tags, they could feed together. But the setup posed a dilemma for the others. If one could feed, its partner was famished. What were the starving birds to do? Would they heed the rumbles of their empty stomachs and abandon their mates? Or, would they rather starve and consort with their partners busy stuffing themselves?

Animals frequent spots with resources, especially food, where they hang out with others that share the same area. One would expect filling their bellies would be a higher priority than bonding with their partners.

However, the great tits surprised the scientists. Instead of going to another feeder to hog on sunflower seeds, they opted to go hungry and stick close to their mates. These couples spent four times longer at feeders than pairs where both partners had ready access to the same feeders.

By accommodating their partners' needs, the birds let themselves in for other changes, such as with whom they socialized. Under normal circumstances, birds would consort with their buddies at feeders. At the number-triggered feeding stations, they had time to kill while their mates gorged themselves. The hungry ones used the opportunity to fraternize with strangers and build new social relationships.

Did the pairs always fly to the same feeder? If not, did males follow the females or vice versa?

Some had no preference. They tailed their mates and were followed equally. In others, one bird of a pair insisted on seeking seeds from its favourite feeder and its mate had no choice but accompany it. More than the birds' sex, the dynamic between the couple decided which one led the way. 'In future work, it'd be great to test what causes some individual birds to take control and dominate the decisions of the pair to suit their own needs,' says Firth.

The researchers suggest mates are a resource, just like food. The birds do their utmost to keep the bond with their partners strong so they can be successful parents. They sacrifice their immediate desires (food) for long-term gains (progeny).

The great tits had another surprise in store. The devoted birds that stayed close to their partners forsaking food didn't starve to death.

Great tits are known for their problem-solving skills. Within a few days, the enterprising birds observed that the feeders' locking mechanism had a two-second delay. As soon as their partner got seeds and moved off, they snuck some out of the feeder before it locked again.

Over the three-month experiment, the ones that were excluded from the containers got better at grabbing food from them. What's remarkable isn't that birds gave up food to stay with their mates. They recognized a loophole in the system and exploited it.

19

Dodging Old Age[1]

COMMUNAL insects such as ants and honeybees assign labour by caste. Workers tend to larvae and queens, forage for food and take care of housekeeping. Soldiers defend the nest against intruders. Queens churn out eggs.

Although they are genetically similar, queens outlive sterile workers. Spending their days within nests protects queens from the vicissitudes of life, while workers brave the big nasty world outdoors. They often die from predation, disease, or environmental exposure.

How do old workers behave? They probably totter on six legs, don't recognize their compatriots as sharply as they did as young ones, and can't smell well – their primary sense. They may also shirk their nursing duties or at least perform them slowly.

Scientists from the US, France and Australia examined how social status affected ageing in *Pheidole dentata*, a kind of big-headed ant found in North America.

No one knows how long dentata workers live in the wild, but they last up to 140 days in laboratories.

The researchers collected ant nests from the wilds of Florida and reared them in the lab. They tracked the ants' ages by monitoring them from the moment they hatched. The scientists tallied how often workers of different age groups carried, fed and cared for the larvae. Whether a worker was young or old, it looked after the offspring with the same attention.

The scientists laid artificial scent trails. By measuring how accurate the ants were, how long they walked, and how much time they took, the researchers assessed the insects' performance.

No matter what the workers' age, they had no trouble following trails with accuracy. In fact, the oldies followed the scented paths for longer distances and were more active than youngsters.

Then the researchers tethered a fruit fly within an enclosure. Would the older ants attack half-heartedly or ferociously? Or would they walk past and ignore it? Elderly workers had no trouble in attacking the poor fly.

The experimenters used molecular techniques to examine cell death in the brains of young and old ants. The ones of advanced years had hardly any dead cells in the area that processes information, learning and memory. Nor did the density of synapses decline, so old ants scored high in cognitive abilities.

In elderly humans, declining levels of two brain chemicals, serotonin and dopamine, causes memory loss, inability to learn, and difficulty in physical movement. However, old ant workers showed no dip in these substances.

What do these results mean? Age has no effect on ant behaviour. Old workers smell as acutely as younger ones. They are limber and strong enough to walk great distances foraging for food. Back in the nest, they are better nurturers of larvae and cleaners of eggs. These workers work until the last weeks of their life.

Is this species unusual in dodging senility? 'We have no reason to believe this ant is unique among social insects,' replies Ysabel Giraldo, who led the study. 'Its social organization is fairly typical of many ant species.' Researchers haven't studied similar insects with which to compare dentata ants.

Since worker ants possess tiny brains, they don't burn up energy storing information. Instead, they process it as a community. Nor do workers reproduce. The authors suggest the insects may channel their vigour to hold off neurological and behavioural decline.

In one study led by Olav Rueppell of the University of North Carolina-Greensboro, ageing honeybees show no deterioration.[2] However, other scientists say the sense of smell in foraging honeybee workers is poorer than ones that work within the hive. Their jobs, more than dotage, cause their downward slide.[3]

Rueppell explains why the two studies seem to contradict. Honeybees don't age as they grow older. Instead, their job sites wear them down. 'This is presumably due to a combination of environmental and internal factors,' he says.

Giraldo points out that despite similarities, honeybees and ants have different social organizations. 'Notably, honeybees progress through tasks as they age,' she says. They start out as nurses, then turn housekeepers, and lastly, foragers.

'Most ants exhibit much more fluid age-based division of labour,' says Giraldo. Senior dentata workers, for instance, assume more tasks. Mature ones, over twenty days old, take care of the brood, forage, and carry out other responsibilities.

Another variation is the mode of transport. The flying insects suffer wing damage after many foraging flights while the situation doesn't arise with the walkers. 'This difference in ecology could lead to differences in ageing patterns,' says Giraldo.

'The clock is not necessarily ticking away for each and every living system,' Rueppell says. 'So, the inevitability of ageing, senescence and death seems a little less inevitable.'

If geriatric workers show no signs of senility, do they suddenly roll over on their backs and breathe their last?

Giraldo doesn't know what causes the death of lab ants as performing necropsies isn't easy. Researchers studied fruit flies for a long time, and they still don't know what kills them. She collected healthy-looking ant colonies from the wild, but they may have had parasites or pathogens.

Ants may not have dodged death, but they seem to have discovered the Fountain of Youth.

20
Shark Navigation[1]

UNDERWATER navigation is complicated. The terrain has no landmarks or paths. Below the surface and in the shallows, the sun may show where east lies, but it provides no more clues.

How do white sharks find their way between California and Hawaii, a distance of over 3,600 kilometres? What map do they follow during their voyage? How do they not wander off course and arrive in Polynesia to the south?

In 2007, Jayne Gardiner and Jelle Atema from Boston University, Massachusetts, suggested smell is the sharks' primary sense organ although they may also rely on vision, water currents, electroreception and touch. Odour itself is not directional. We nail the location of a scent by comparing varying concentrations in the air. Fish depend upon another sense organ called the lateral line to pinpoint their prey's position, said the researchers.[2] It is a series of electrosensory organs located under the skin on the snout and along either side of the sharks' body.

It is one thing for the fish to sniff down delectable morsels close by, but another to trace a coast thousands of miles away.

In some species, the olfactory bulb, a part of the brain that decodes scents, doesn't scale in proportion to body size. It is outsized in far-ranging tiger and white sharks compared to sedentary reef sharks.[3]

Based on this observation, in 2012, neuroscientist Lucia Jacobs speculated about the primary function of smell.[4] 'Instead of the main olfactory system simply telling the brain "what" something is, I proposed it is telling the brain both "what" and "where",' says Jacobs. 'An odour plume, whether in air or water, can act as a sensory highway, which an animal can use to calibrate its navigation cues from other senses, such as vision. In water, however, vision is only short-range. So, for long-distance navigation, I proposed that olfaction could be a very important method for navigation.'

Andrew Nosal from the University of California, San Diego, and his colleagues conducted an experiment to test this hypothesis. They chose the leopard shark (*Triakis semifasciata*), a species with large olfactory bulbs, which congregates off the Pacific coast of North America.

At an average length of a metre and a half, leopard sharks are no danger to humans. They stick close to shore, along the bottom of muddy or sandy flats, kelp beds and rocky reefs, but they also make occasional forays into the open sea.

Nosal and his team caught eleven female leopard sharks. They hung a strong neodymium ring magnet five centimetres above the bottom of the holding tank to garble the fish's magnetic reception. If they changed the water in the tank to aerate it, the sharks could pick up chemical cues. Instead, they used compressed air. The helmsman spun the boat in figure of eights to disorient the creatures before heading northwest to the release site. An opaque tarpaulin covered the container so the fish couldn't see where they were going.

The researchers weren't done with the sharks. They plugged their noses with petroleum jelly-soaked cotton wool balls. Sharks don't breathe through their noses, so the plugs can't suffocate them. For the twenty-four hours the wads take to disintegrate, they prevent the fish from smelling. Before letting them go, the researchers tagged them with acoustic transmitters with timers set to detach and float in four hours.

Nosal and team released the leopard sharks nine kilometres offshore. The open sea is a hostile zone for these creatures of the coast. It is not a place to hunt clams, crabs and fish, and there are many predators abroad.

Until the transmitters disengaged, these sharks had made on an average a third of their way to shore. They swam slowly in convoluted paths, seemingly lost.

Nosal's team caught another fifteen sharks and put them through the same treatment – the magnet in the tarp-covered holding tanks, figure of eight spins, the works. Except for one thing: they didn't plug their noses. The researchers called these sham-controls. The behaviour of these fish offered a useful comparison to the disoriented ones with stuffed noses.

As Nosal expected, the tracks of these sharks ended two-thirds of the way closer to shore in four hours.

What guided the sharks to the coast?

The researchers don't know for sure. Nosal and his colleagues suggest amino acids released by coastal plankton may act like beacons, guiding shore-bound leopard sharks.

Adult salmon that live in the sea probably smell these same organic compounds to find their way to the freshwater streams where they were spawned.

Another fragrant candidate is dimethyl sulphide, a compound produced by phytoplankton and other marine algae, used by seabirds, marine mammals and plankton-eating sharks to find their way.

River mouths, rocky shoreline, reefs and kelp forests release different combinations of compounds. Every location along the coast may have its own signature smell that sharks use to return home.

How do they navigate by recalling the smell of their destination?

'When we released the sharks from shore, two things could have happened,' says Nosal. They may have perceived the faint scent of an odorant that told them they were far from shore. They may have used other senses to find their way back. 'In this case, olfaction is only important when the shark is first released.'

Another possibility is the fish headed in the direction from where they picked up the strongest odours. If they swam away from the coast, the concentration of scents reaching them would be weak, and they'd know they were heading the wrong way.

'We actually observed this in several of the sham-control sharks,' says Nosal. Soon after they were released, they headed in the opposite direction from shore. But in less than thirty minutes, they made a sharp U-turn and corrected course.

Besides odours, sound may aid navigation. The low-frequency pounding of surf could indicate the shore. At the experiment location in La Jolla, surf noise ranges from fifty to 300 hertz, well within the hearing range of sharks. They swam near the bottom where sound travels slowest and farthest, giving this theory credence.

In 2014, Atema and his colleagues from Florida, US, suggested that when specific senses are disabled, the fish successfully identify, track and capture prey by switching to alternate sensory systems.[5]

Nosal speculates that the sharks with stuffed noses eventually found their way to shore using other cues. The sun, for instance, may provide directional information. But some leopard sharks appeared lost on sunny days while others navigated successfully when it was overcast or foggy.

Nosal and his team conducted more experiments. They picked up another ten and dropped them nineteen kilometres away, farther from the shore than the others, without messing with their senses. Since these sharks had numerous cues, the researchers expected them to make a beeline for the coast. But only six reached the shore in four hours. Surprisingly, the rest appeared lost and moved in tortuous paths as if they had stuffed noses. The researchers wonder if only experienced sharks are able to navigate accurately.

If 40 per cent that had use of all their senses lost their way, is the case for olfactory navigation robust?

'There will always be outliers in data sets, which is why we average the results from individuals within groups and use statistics to compare them,' says Nosal. The location being farther from shore, these lost sharks perhaps did not detect the faint and patchy chemical cues emanating from the coast.

Navigating nineteen kilometres ought to be no big problem for a species that crosses the San Pedro Channel, separating mainland California and Santa Catalina Island. The channel is thirty-two kilometres wide at its narrowest point and approximately 800 metres deep.

The researchers say they haven't gotten to the bottom of how sharks find their way, but this is only the first study of wild sharks to explore if the sense of smell plays a role in open-sea navigation.

Atema, who wasn't involved in this experiment, says, 'Based on our own larval reef fish and shark work, I (and others) find it hard to believe that there can be a chemical gradient useful for navigation over nine kilometres with a predominantly longshore current. Currently, there is no evidence for an odour plume in that area of the ocean. However, that does not mean there is no useful chemical information. It means we now need to find evidence for or against such a spatial gradient. And that is exciting. I can think of a number of possibilities.'

Nosal says he plans to determine the exact mechanism for navigation in the future, including what other senses are involved and how they work in tandem. He is in for the long haul since this is one secret that animals and birds have held close despite many years of research.

21

A Parasite that Meddles with Behaviour?[1]

THE protozoan parasite *Toxoplasma gondii* offers rats on a platter to cats. It engineers the rodents' brains so the animals lose fear of their nemesis. Even more diabolically, it seems to make its hosts sexually attracted to the predators.[2] Such rats frolic in areas that reek of feline urine, enticing a hungry cat to pounce on them. Toxoplasma changes the behaviour of its hosts so it can jump into cats' bodies to have sex and produce millions of thick-walled eggs called oocysts. Although it infects a whole range of birds and animals, including humans, it reproduces sexually only inside felines. While the sated cat looks for a quiet place to sleep off the rigours of hunting, the parasite slides into its gastrointestinal tract to meet others of its kind.

A twenty-gram faeces of an infected cat has between two million and twenty million oocysts that lie dormant in the soil for more than a year,[3] waiting for a foraging bird or rodent to pick them up.

Brain-garbled rats don't throw themselves in front of people, nor do most humans eat rodents. People become infected by handling cat litter, but most get the parasite by eating uncooked meat or not washing their hands thoroughly after gardening. The protozoan affects up to 60 per cent of humans.[4] One study said pregnant Indian women with this infection ranged from 9 to 37 per cent.[5]

The studies on the effects of toxo on humans don't stress direct cause and effect. Instead, they look at broad patterns of behaviour among the infected, suggesting the parasite affects our brains too. It delays people's reaction times and reduces their ability to concentrate. Infected men were suspicious, dogmatic and unlikely to heed the rules of society. Affected women were warm-hearted, extroverted, and easy-going.[6]

Why does toxo change our behaviour[7]?

According to one theory, these alterations are inadvertent side effects of infection. To escape the onslaught of the host's immune system,[8] toxo causes cysts to form in the brain, which may change the behaviour of the host. Another hypothesis says the parasite manipulates our personalities without realizing we are no longer on any cat's menu.

To understand the origins of toxo's manipulation of humans, researchers from the Centre d'Écologie Fonctionnelle et Évolutive, Montpellier, France, tested our close relatives, chimpanzees. Would the primates infected by the parasite react to the urine of large cats?

By subjecting thirty-three chimpanzees at a captive facility in Gabon to an antibody test, the researchers diagnosed nine were infected with toxo. They presented all the primates that live in five groups with urine samples of humans, leopards and tigers or lions, one at a time in a random order. Each group had twenty minutes to investigate before the piss was washed away.

The researchers analysed how often chimps approached the urine, whether they smelled it by sticking their noses into it, if they sniffed their hands after wetting them first, or if they tasted it with their tongues. Every thirty seconds, they noted which individuals were in a one-metre radius of the sample.

Toxo-infected animals were typically bolder, three times more likely to investigate leopard urine than uninfected ones. With the same ease as it altered rats, the parasite meddled with the primates' sense of fear.

If wild toxo-infected chimpanzees lurked around spots where leopards spray their territorial boundaries with piss, the cats could slay them, becoming the breeding grounds for toxoplasma.

Infected and uninfected chimpanzees were not interested in the odour of lion and tiger urine. The authors explain these cats are not chimpanzee predators. Lions live in open savannas while chimpanzees are creatures of the rainforests. Tigers are Asian while chimpanzees are African. The researchers say the parasite specifically manipulates its host to fall prey to their natural cat predators – leopards.

Since infected chimpanzees behaved similarly to toxo-infested rats, the authors speculate the parasite causes the change in the primates' behaviour too. Do such chimps get eaten by leopards more often than healthy ones? Unlike in the case of rodents, it is impossible to prove, say the authors.

The study raises two issues: how does toxoplasma, which lives in many species, know which animal it has affected and what that animal's natural predators are? Could it tell the difference between lions, tigers and leopards?

No researcher has been able to come up with a mechanism that reliably separates two related predators in the brain of prey, says Ajai Vyas, a neurobiologist at Nanyang Technological University, Singapore. He suggests the differences in reaction to the urine of leopards and other cats is probably because of the way the experiment was designed.

'An alternative explanation is that x ml of predator A urine is not perceptively equal to x ml of predator B urine,' says Vyas. Perhaps, some samples had a weaker stench. 'So, the researchers might be measuring different doses of predator odours when they think they are measuring two odours.'

Secondly, the authors cite a 1991 paper that estimates an individual chimpanzee's risk of being eaten by a leopard at 30 per cent per year in Côte d'Ivoire.[9] A more recent study, published in 2002, conducted in the same forest shows leopards don't prey on chimpanzees. Only one in 200 leopard faeces had chimpanzee remains.[10] Neither is the case of lion predation settled. In 1989, two lions preyed on four chimpanzees in Tanzania's Mahale Mountains National Park, one of the few sites where the two species share the same habitat. Other troop members reacted to the lions by alarm calling, whimpering, climbing into trees or being silent.[11] This may be a rare event, but it may happen more often than we realize. In sum, the selective attraction of leopard urine alone to infected chimps isn't convincing.

What does this say about the parasite's effects on humans? The authors suggest it could cause behavioural changes in humans because their ancestors were cat prey once. Did toxo evolve with a common ancestor of humans and chimpanzees? Even if it did, Vyas points out that many of the current strains are of recent origin. Nor is there evidence yet of independent evolution after the ancestral line split into the two primates.

Another possibility is that the brains of diverse species, including rodents and primates, react to the infection in similar ways. '[This] option seems the most plausible to me, and that contradicts the idea that chimps will show a specific change only to leopard, and not to other feline predators,' says Vyas.

Although research on how and why the parasite influences human behaviour is ongoing, this study indicates the challenges facing researchers.

Contradicting previous research, a more recent study[12] says infected rodents aren't fatally attracted to felines alone. The normally shy and timid animals become bolder and fearless of threats in general.

The only animals to escape toxo's fiddling are cats. While severe infestation affects their health, most cats don't seem to suffer any ill-effects. The parasite offers one advantage to felines: it makes hunting much easier.

22

High Speed Bird Vision[1]

SMALL birds, when pursued by raptors, zip through the woods, ducking branches and swerving around tree trunks. Flycatchers dart after flying insects with astounding speed and accuracy. Not only do they match their prey's acrobatics without suffering motion blur, they predict the flight path of their prey, and nab them at a point in space where the insect-eaters expect they will be. Rarely do these birds fly out of control and crash headfirst like speeding motorists. Nor do they collide with obstacles.

Researchers say high metabolic rates allow photoreceptors in the eyes of small, agile birds to react rapidly.[2] They measure the number of changes birds can perceive per second to gauge this ability called temporal resolution, which works best in bright sunlight.

Scientists from Uppsala University, Stockholm University and the Swedish University of Agricultural Sciences predicted blue tits (*Cyanistes caeruleus*), collared flycatchers (*Ficedula albicollis*), and pied flycatchers (*Ficedula hypoleuca*) could see changes up to 100 hertz. All three belong to a group called passerines or perching birds.

Flycatchers make ideal experiment subjects because they outmanoeuvre their insect prey. Blue tits are insectivorous only during the breeding season, but the researchers included them in the study because Stockholm University's Tovetorp research station in Södermanland had an ongoing research project on them.

They caught eighteen birds of the three species from the wilds of south-eastern Sweden and Öland island in the Baltic Sea. They trained the captives to find rewards placed under a flickering lamp – blue tits got peanuts and the flycatchers mealworms.

When given a choice between blinking and steady lights, the birds made for the former. Once their subjects fetched a treat at least four out of five times, the researchers incrementally increased the speed of the flashing until the experimental subjects could no longer decide which lamp flashed.

This threshold, called the critical flicker fusion rate, for blue tits was 131 hertz and collared flycatchers at 141 hertz. The pied flycatchers' was a high 146 hertz. On average, the visual acuity of the three species is about 40 hertz above the highest rate for any other vertebrate. Their vision was so fast that Usain Bolt's 100-metre dash would seem like a slow-motion stroll. In comparison, the critical flicker fusion threshold for humans is a mere 60 hertz. The birds' performance far exceeded the researchers' expectations.

The results surprised other scientists too. Such fast vision is comparable to insects rather than vertebrates, says Kevin Healy, a research fellow at Trinity College, Ireland. 'It will be very exciting to see how other species with different hunting strategies might fit into the overall picture.'

'Before this study, most of what we knew about avian temporal resolution came from studies of domesticated, ground-dwelling birds or those that live in rather dim or dark environments,' explains Anders Ödeen of Uppsala University, who headed the experiment.

The results have implications for captive birds' welfare, he says. His earlier work with chickens' flicker rate, an average of 87 hertz,[3] highlighted these concerns.[4] Many low-energy light bulbs, fluorescent lamps, and LED lights flicker at 100 hertz. Even if the flicker is invisible to our eyes, captive songbirds, with vision more than twice as fast as ours, are likely to see it as a flashing strobe that can cause stress, behavioural disturbances and discomfort.

Do all small birds have rapid vision?

In a subsequent study,[5] Ödeen and other scientists investigated the temporal resolution of budgerigars (*Melopsittacus undulatus*). These popular pets belong to the parrot family, distantly related to the passerines. In the deserts of Australia, the seed and grain-eaters neither perform flying stunts to catch a meal nor escape predators by zooming through the obstacle course that is a forest.

The budgies' visual acuity, between 77 and 93 hertz, was comparable to domestic chicken, but much less than tits and flycatchers. It is likely they aren't affected by our lighting systems, but the researchers caution that more studies are needed.

More than their small size and high metabolism, birds' lifestyles seem to drive the speed of their vision. Budgies, living in open country and feasting on stationary seeds, don't require an ability to notice changes in microseconds. The wild red jungle fowl of Asia, from which the chicken was domesticated, prefer rooting around in the dark forest floor and walking to flying. However, diving after insects and surviving high-speed chases through a thicket of tree trunks may put acute pressure on flycatchers' eyes to become faster at seeing obstacles.

Fast vision comes at a price as these birds don't have keen eyesight. Birds of prey, like eagles, have the sharpest vision known, capable of spotting quarry on the ground while they hover high in the sky. Their ability to track a number of details per degree in their field of vision, called spatial resolution, ensures they cannot keep up with flycatchers in full flight.

Flycatchers, however, don't get a clear image of their pursuer. A blurred one is all they need to zoom out of harm's way and to see obstacles in their escape course.

23
Empathetic Rodents[1]

SCIENTISTS thought for a long time that empathy was a uniquely human trait. But in recent years, many species such as great apes, dogs, crows and elephants demonstrated emotional understanding of others, leading experts to think empathy needed high intelligence.

When an animal licks, grooms or touches a miserable individual, it is consoling,[2] say primatologists Frans de Waal and Angeline van Roosmalen. And this indicates compassion.[3]

Even though rats don't display the level of intelligence displayed by chimpanzees and elephants, they are capable of empathy. A study showed rats resisted the temptation to eat chocolate and instead went to rescue others treading water in an adjoining compartment.[4] The only way the wet, miserable ones could get out was if another rat opened a door. Rodents that had been in the shallow pool before were quicker to sense their mates' distress. Since they aided their companions 50 per cent to 80 per cent of the time before nibbling on the treat, their desire to help was at least as strong as the motivation for food.

Prairie voles (*Microtus ochrogaster*) also comfort their friends and companions. These brown mouse-like rodents are among the few life-long monogamous animals (see Chapter 16: Rodent Monogamy). Since they cooperate and show concern for each other, scientists from Emory University, Georgia, US, suspected the voles might respond to others' distress.

The researchers separated pairs of voles that were mates, siblings or long-term cage-companions. They gave a mild electric shock to the feet of one group, the 'demonstrators'. Although the other group didn't see their partners getting zapped, they were called 'observers'. The two groups were then reunited.

The observer voles had no idea what had happened to their mates, but they immediately began licking their buddies and didn't stop for up to ten minutes.

When the pairs united without receiving any shocks, the observer voles didn't spend as much time grooming their buddies. The rodents gauged whether their partners were stressed and in need of comforting by merely looking at them.

Consolation helped the agitated voles become normal again and kept the level of corticosterone, the stress hormone, in their blood from surging.

When researchers didn't allow traumatized voles to reunite with their mates, the animals remained scared and upset. Observing them through a clear perforated sheet, their companions also grew anxious and had higher levels of corticosterone. Viscerally mimicking another's emotional state indicates empathy, say the authors.

The research team notes that voles commiserated only with known individuals. When they saw unfamiliar ones with high stress levels, they didn't rush to console them. Humans are similarly biased too.

We feel compassion when the hormone oxytocin kicks in. The chemical is linked to maternal care and pair-bonding, and it also enables us to read another's emotional state.[5] It acts on a particular part of the brain called the anterior cingulate cortex. This area not only regulates blood pressure and heart rate but is also vital for empathy.[6]

Could closely related meadow voles also have the trait? In experiments, they weren't considerate to their stressed peers.

'Unlike meadow voles, prairie voles have a considerable amount of oxytocin receptor, a sensor for oxytocin, in key emotional brain regions crucial for bond formation,' says one of the researchers, the neuroscientist Elissar Andari, who studies autism and neurobiology of human social behaviour along with animal research.

When the scientists gave a shot of oxytocin-blocker to the observers' anterior cingulate cortex, the prairie voles showed no concern for their partners. This led to the conclusion that the interaction of oxytocin with the cingulate cortex was essential for empathy.

If prairie voles can console, then the origin of compassion must go farther back in evolution. It is likely to be more common among animals, and it doesn't need any great cognitive abilities. Why is empathy so hard to find then? The authors say consolation 'emerges only under particular social and evolutionary conditions'.

It may also come at a cost. By soothing companions, one not only sacrifices one's own happiness, but also invests time and energy in providing assistance, says the study's main collaborator, James Burkett. 'However, highly social species, like humans and prairie voles, can also benefit from showing empathy and comforting others, in terms of peace, cooperation and community. In this way, consoling can be considered an altruistic behaviour.'

If species as disparate as humans and prairie voles share a common evolutionary heritage, could meadow voles and other rodents empathize but are unable to express it?

'It is possible that other rodents have other forms of expression that can be more subtle and less explicit to the human eye,' says Andari. 'Hence, it would be more difficult to observe or measure.'

Scientists apply such animal studies to help humans with developmental disorders, such as autism, and mental health issues, including schizophrenia and psychopathy.

When Andari[7] gave intranasal oxytocin to autistic patients, they paid more attention to social cues. 'By studying consolation in prairie voles,' she says, 'we can understand the neural circuitry and unravel the role of molecules such as oxytocin in triggering such behaviour.'

Until researchers find other animals that commiserate, albino lab rats and prairie voles remain the odd ones to possess what was once thought to be a uniquely human trait.

24

Bizarre Spider Sex[1]

THE Darwin's bark spider (*Caerostris darwini*) has one of the most interesting sex lives in the animal kingdom.

Scientists discovered this native of Madagascar in 2010[2] and named it in honour of Charles Darwin on the 150th anniversary of his work *On the Origin of Species*. Since then, they have realized this is one extraordinary spider. The females spin the world's largest web, covering up to 2.8 square metres, an area larger than a 100-inch flat-screen television. The orbs' anchor lines stretch up to twenty-five metres across rivers and streams. Their silk is the toughest biomaterial in the world, ten times stronger than bullet-proof Kevlar.[3] On this sturdy silken hammock, the spiders engage in bizarre sex.

The females are fourteen times heavier and more than double the size of males. This difference between the genders, called sexual dimorphism, is common among spiders. But the females are far from monstrous; they are only two centimetres long. Tiny suitors vie to mate with a large mamma at great personal cost. Like females of many other spider species, she could emasculate, or worse, eat them. Why they do this is still unclear. Theories range from nutrition to sexual selection.

Researchers from the Slovenian Academy of Sciences and Arts, Ljubljana, noticed the males prefer to mate with freshly moulted members of the opposite sex as they are soft, calm and unable to attack. The arachnids copulate without maiming or loss of life. But once their skins harden, females turn predatory. To protect themselves, males wrap their mates with silk before hazarding coitus.

Mate binding slows down female attacks but does not prevent them, says Matjaž Gregorič, the chief collaborator in the study of the sexual habits of the species. At least it increases the males' chances of survival. How could a spider trussed up with the world's toughest biomaterial break free?

They don't bind their mates like a mummy, says Gregorič. The females can extricate themselves without effort. The males don't bother with this S&M manoeuvre with defenceless, freshly moulted partners.

Sex between spiders is complicated business. Males exude the sperm on to a specially constructed sperm web. Using a pair of leggy special appendages near the mouth called pedipalps, they scoop it up and insert it into an opening called epigyne high in the females' abdomen.

While mating, male Darwin's bark spiders may damage their palps. In half the sexual encounters the researchers recorded, the organs broke off inside the females where they could act as genital plugs to prevent the latter from copulating again. This is a common strategy in spider species with disproportionately small males. In another species of orb-weaving spider,[4] scientists found the severed male bits continued to transfer sperm.

However, evolution favours female Darwin's bark spiders by rendering the plugs ineffective. Later suitors remove them, freeing the females for more sex.

Since detaching parts of their organs doesn't prevent females from breeding with their rivals, the males take their emasculation further. Within twenty-four hours of copulation, they chew off what remains of their maimed palps. Unable to use their crippled organs any more, they have little to lose and everything to gain.

'This [removal of palps] results in hormonal changes that render eunuch males more aggressive, and such males are good at fighting off rival intact males,' says Gregorič.

By transforming into fierce combatants, they guard their partners against rivals and prevent them from becoming impregnated again. By force of circumstance, males of the species are monogynous, having sex with one female. Occasionally, they may mate with two but not more.

None of these behaviours – self-mutilation, cannibalism, bondage, genital plugging – surprised the researchers as they were aware of spiders' bizarre sexual acts.

What they didn't expect was the oral sex. The males hook their jaws, or chelicerae as spider mouthparts are called, to the females' epigyne and salivate.

Since the fluid excretes from the mouth, 'we simply call it saliva,' says Gregorič. 'At this time, we cannot know for sure if venom or other glands contribute to this liquid.'

Irrespective of the females' age, condition, and aggressiveness, the males drool over the epigyne. Nor do they restrict this performance to foreplay alone. They go at it between bouts and even after copulation as many as 100 times in a session. All male spiders in the study performed oral sex.

The behaviour is not common. Some mammals such as bonobos, hyenas and dolphins practice fellatio. But cunnilingus is rare. Researchers know of only bonobos, Indian flying fox and fruit flies engaging in it until now.

'I hypothesize this behaviour to be more widespread in the spider world, perhaps connected to sexually size dimorphic spiders,' says Gregorič. He points to a 115-year-old forgotten text, *Apareamiento del Latrodectus formidabilis,* describing oral sexual contact but not salivation in black widow spiders.

Why do these male arachnids engage in this bizarre act? Do they avoid being dinner? The Slovenian researchers speculate this is unlikely since males salivate with freshly moulted females that don't pose a threat. Nor does the saliva seem to remove genital plugs from an earlier mating because they drool during copulation and afterwards.

The scientists say the males may use oral sex to boost their chance of fathering the next generation. Enzymes in the saliva could aid in sperm transfer and stymie rivals' success. But they need to test these speculations.

Despite the males' attempts to protect themselves and perpetuate their genes, 31 per cent of the females cannibalized their partners.

25

How Do Echidnas Survive Wildfires?[1]

EVERY year, wildfires sweep across swathes of Australia. How do small, slow terrestrial creatures survive?

Australian biologists from the universities of New England and Western Australia and Curtin University, led by Julia Nowack, investigated by implanting temperature data loggers and sticking radio transmitters on ten adult echidnas.

Short-beaked echidnas (*Tachyglossus aculeatus*) belong to a group of egg-laying mammals called monotremes. These hedgehog-like animals colonized most of Australia, from arid deserts to cool rainforests, where they rip into rotten logs, termite mounds and ant nests with their long claws during the day. When the heat of summer gets to them, they go into torpor, a state similar to hibernation, and emerge only at nights. Such dormancy lasts from a few hours a day to a few days.[2]

The Department of Parks and Wildlife conducts prescribed burns to reduce the severity and extent of wildfires. On 21 April 2015, it burned 227 hectares of Dryandra Woodland, south-east of Perth. Five of the study's echidnas lived in the burn section, and five were outside. Comparing the two groups allowed the biologists to study behaviour and fluctuations in body temperature before, during and after the burn.

Bushes and trees burned, and timber smouldered for weeks, destroying eight of the fourteen logs that echidnas used as hideouts. One of the tagged animals died when the hollow log in which it was resting ignited. Two other untagged echidnas suffered the same fate.

Were these animals in such a deep slumber that they didn't sense the fire in time? The researchers say the tagged echidna's body temperature was just above 23°C, not so cold it couldn't move.

'I really have no explanation why he did not flee, except that he "made the wrong call",' says Nowack.

Other mammals such as mouse-like dunnarts woke up on smelling smoke.[3] Indeed, another echidna that shared the log with the roasted one aroused from torpor and escaped to shelter 160 metres away.

The survivors holed up during the day and restricted their activities even at night, for at least three weeks after the fire. Three went into a deep sleep for five days. Echidnas outside the fire zone, however, behaved normally, going about their daily business.

Despite the scorched earth, none of the monotremes left and sought refuge elsewhere. Not even the two animals whose territories burnt completely felt the need to move. Instead, the prickly creatures decamped to unburnt stumps. Deep nonflammable burrows would make a better choice.

In another study, echidnas retreated to cavities and underground chambers for long bouts of hibernation.[4] They ducked into logs only during the hottest time of the day. So why did the monotremes prefer fallen trunk boles?

The unfortunate ones living in the fire zone didn't have a choice of refuge. The area had many massive stumps but no rocky hollows or tunnels.

Burning wood and smoke may not send the monotremes running for cover. Some rested away from the scorched region but later ventured into it even though it was still blazing. 'The echidnas did not seem to avoid the burnt areas of their territory at all,' says Nowack, 'and we found echidnas inside heavily burnt patches roaming around, their noses full of ash.'

The fire destroyed wood and leaf litter that would have teemed with ants and termites. During a blaze, these insects retreat deep into subterranean chambers and out of reach of their predators. Torpor is a useful strategy to wait out food scarcity after a conflagration.

Since their metabolism shrinks, sleeping saves them the trouble of filling their bellies[5], with the added advantage that they can avoid predators.

When the monotremes emerge, they have to work hard, walking farther and searching harder. Although the animals in the charred section chilled out for up to five days, they can last in this state of limbo for months.

Fire isn't the only natural disaster that mammals like echidnas sleep through. They also slumber through other adversities such as droughts and floods. Sugar gliders, for instance, switch off during cyclones.[6]

The researchers suggest ancient ancestors of mammals may have survived a cataclysmic event by going into torpor. About sixty-six million years ago, something killed almost all large terrestrial creatures. Most scientists suspect an enormous asteroid estimated to be 10 kilometres in diameter slammed into the earth. During this cataclysmic event that they call Cretaceous–Paleogene (K–Pg) boundary, burning fragments or an infrared wave from the blast would have set off fires around the world that lasted for a year or more. Life on our planet was never the same again. With large fauna out of the way, the surviving small mammals evolved and occupied the vacant niches.

These mammal ancestors may have hunkered in subterranean burrows and gone into torpor, proposed Barry Lovegrove and his colleagues at the University of KwaZulu-Natal, South Africa.[7]

Lovegrove and his team cite the example of the common tenrec (*Tenrec ecaudatus*), an opossum-like mammal, which goes into an eight to nine-month-long hibernation during Madagascar's mild winter.

If mammal survivors of the meteor strike were like these echidnas,[8] they would have found a good deep burrow and gone dormant. When environmental conditions became better, their descendants flourished.

Almost half the mammal extinctions worldwide over the past 200 years occurred in Australia. Of these, more than 90 per cent were animals that, like humans, maintain a constant body temperature. Creatures that go into torpor made up less than 7 per cent of the species that disappeared.[9] Of the existing ones, at least 43 per cent of the native mammals in this fire-prone continent go into torpor, suggesting that sleeping it off is a survival strategy.

'Even though our study can obviously not prove that torpor facilitated mammal survival after the meteoroid impact at the K-Pg boundary,' says Nowack, 'seeing the response of echidnas to the fire makes me strongly believe that torpor would have played an important role for survival.'

According to predictions, climate change will increase the frequency and intensity of natural disasters. The species to master slumber may be the survivors.

26

Plants with Private Armies[1]

PLANTS can't run away to escape from herbivores. But neither are they helpless victims. Physical and chemical defences such as sharp thorns and alkaloids are common. But some go further and recruit defenders. The trees of bullhorn acacia of Mexico and whistling thorn of east Africa enlist an army of ants by providing barracks in their thick hollow thorns. At least 4,000 species, such as lima bean vines, exude nectar from special glands called extrafloral nectaries. The rich liquid is manna for the insects with a sweet tooth. Whether the plants provide room or board, their insect tenants come out in force at the slightest disturbance, ready to do battle to protect their homes – the plants.

The European bittersweet nightshade (*Solanum dulcamara*) has an unusual modification to this strategy. It produces toxins in its bright red berries, tubers and flowers. But its green arrowhead-shaped leaves are fair game for pests, such as flea beetles and slugs.

Anke Stepphuhn, at the Freie Universität Berlin, Germany, and her Ph.D. student Tobias Lortzing let caterpillars feed on the nightshade vines in a greenhouse for another research project. The climbers oozed sweet, sticky secretions from the wounds, surprising the researchers. Typically, plants close their injuries quickly to prevent infections and loss of valuable resources. The scientists remembered seeing ants swarming over the creepers in the wild and realized there was a connection.

They cut the nightshades' leaves with a scalpel, and the vines secreted some sweet fluid. When they damaged the foliage repeatedly or coated the blade with jasmonic acid, the vines bled profusely. Jasmonic acid, a hormone exuded by distressed plants, induces the secretion of sugary liquid in extrafloral nectaries. Acting as alarm bells, the amber beads of syrup summoned three species of ants to the climbers' aid. The insects patrolled injured leaves and plants longer than undamaged ones. Not only did the creepers alert their insect soldiers of the attack, but they also showed where the battlefield lay.

In other ant-recruiting plants, specialized tissues create the sweet liquid and specific structures in particular locations dispense it. However, the nightshade secretes nectar anywhere it is hurt – leaves, petioles and stems. Every wound becomes a nectary.

Both larvae and adults of a species of flea beetle are a common pest through the growing season. The grubs hatch from eggs in the soil and crawl up the vines. Once they burrow into the pith of the stem, the plants can do nothing about them.

The researchers observed the creepers secrete nectar and summon their ant hordes to remove the worms before they could dig themselves in. The biting insects killed a third of the pests and reduced damage to the plants.

After the larvae matured into beetles, they emerged from the stems and took small round bites of the leaves, causing the distressed nightshade vines to secrete wound nectar. Climbers under attack had three times more ants than undamaged ones.

A fortnight later, the affected plants had grown and put forth more shoots.

Steppuhn and her team analysed the chemical composition of wound nectar and normal plant sap. While sucrose is the main ingredient in both, wound secretions have no other sugars. Instead, they have three amino acids and an antioxidant ascorbic acid, commonly found in flower nectars.

The researchers sprinkled sucrose on some vines and plain water on others. Black slugs, another plant pest, didn't have as much success in feeding on the sucrose-treated plants. The ants attacked and killed many, and the survivors lost 20 per cent of their body mass. Plants thus protected by ants suffered only half the damage of ones sprayed with plain water.

The health of the climbers depends on the insects.

In real life, however, the vines don't secrete much fluid from the damaged edges when black slugs chew up nightshade leaves.

'They eat too fast and often the whole leaf,' say Lorting and Stepphuhn. 'There is no tissue left to secrete nectar.'

This extraordinary defence, previously unknown in the plant world, generated excitement among researchers.

'It is one of those discoveries that plant ecologists will think "how could I have never observed this?"' says Anurag Agrawal of Cornell University, US. 'This research group has found perhaps the simplest mode of wound-inducible extrafloral nectaries, by letting herbivores open the tap.'

Vartika Mathur of Sri Venkateswara College, New Delhi, points out the advantage of not possessing exclusive secreting and dispensing spots for sucrose. The plants do not waste resources on extrafloral nectaries when there are no herbivores.

Experts speculate the simplicity of this process may be an intermediate step in the evolution of dedicated glands that help plants maintain their own personal armies.

Lorting and Stepphuhn suspect other plants that produce juice from injuries may have so far escaped attention.

'This wound nectar is really hard to find under natural conditions,' they say, 'because it is either quickly consumed by animals, dries to small sugar residues, or is washed away by rain or morning dew. So, it might be a largely overlooked phenomenon because nobody was looking for it.'

Another team conducted a survey of an evergreen broadleaved forest in southeast China.[2] It found twenty-two species of ants crawling over chewed up leaves of twenty-three of forty trees examined. The researchers observed the insects licking up sap from the wounds. Until now they didn't know these trees produce nectar anywhere besides their flowers.

How do these trees and plants make the fluid with no devoted glands? That remains unanswered.

The ancients, who named the plant bittersweet nightshade, seemed to be aware of its arsenal of vile toxins and sweet nectar that sends its predators packing.

27

Did Moby Dick Sink the Pequod?[1]

COULD a fifty-six-ton Moby Dick ram and sink Ahab's 240-ton *Pequod*? Since 1796,[2] whalers claimed sperm whales (*Physeter macrocephalus*), which they hunted in the thousands, deliberately sank their ships by colliding with them.

Owen Chase wrote the first detailed document in 1820, when he recounted his experiences as the first mate aboard the 238-ton whaler *Essex*. A monstrous twenty-six-metre cachalot (meaning 'big head' in Spanish and Portuguese), as the species was also called then, 'distracted with rage and fury' pummelled the vessel. Although the ship's copper-plated hull was built of white oak, one of the strongest woods available then, it sank in the South Pacific. This extraordinary bull inspired Herman Melville's *Moby Dick*. Chase proposed the unique design of these giants' heads made them battering rams. How reliable are these old accounts?

Males of many species grow larger than females, a phenomenon called sexual dimorphism. They deploy horns or other appendages to beat up their opponents and win mates. For instance, blackbucks use a prominent set of ringed corkscrew-like horns to defend their harem of tan-coloured, rackless does from rivals.

Males of other cetaceans such as humpback whales, narwhals and killer whales ram each other. Male sperm whales are three times larger than females and may put their enormous noses, the world's largest, to good use. However, reports of these giants butting each other are rare.

On 30 January 1997, wildlife pilot Sandra Lanham spotted two male cachalots as she flew over the Gulf of California, Mexico. She estimated they were more than six kilometre apart and swimming toward each other at 17 kilometres per hour. They dove below the surface before colliding head-on with each other. 'The forehead of the sperm whale is one of the strangest structures in the animal kingdom,' says Olga Panagiotopoulou, University of Queensland, Australia, and the main collaborator in the study. Citing this eyewitness account, she suggests the cetaceans smash heads to overpower rivals.

Is this a common behaviour?

'There is no physical reason sperm whales could not butt each other in aggressive or non-aggressive encounters,' says Luke Rendell, University of St Andrews, UK. 'However, aside from the single account offered by the pilot in the paper, there is no report of this behaviour from any of the long periods of time researchers have spent with the animals. Hal Whitehead [Dalhousie University, Canada] describes an account of one whale grasping the tail stock of another in its jaws, while I have personally observed two males swimming toward each other blasting each other with very rapid and very loud clicks – but no butting.'

Biologists are sceptical because sperm whales' heads house sensitive communication and navigation organs. Plunging to over 1,000 metres, these deep-divers hunt giant squids in the dark by echolocation They make click sounds through a pair of lip-like valves in the head, which locate prey and also communicate with others in the pod. Different populations have distinct dialects, and each sperm whale has its own unique accent or pattern of clicks.

Besides their echolocation gear, two enormous oil and wax-filled organs – the spermaceti organ and the junk sac – take up most of the space in their gigantic noses. From the early eighteenth century, whaling ships targeted sperm whales for their spermaceti. Oil drained from their heads not only lit the streets of London, New York and Paris, it was a key ingredient in candles, soaps and lubricants. Whalers thought the oil was semen and hence the name 'spermaceti'. Cachalots use these glands to direct and amplify their sonar clicks, stunning their prey with a high decibel buzz.

If these cetaceans used their heads to batter each other, they could cause irreparable damage to these sensitive communication, stunning, and navigation structures.

Back in 2002, David Carrier, at the University of Utah, US, suspected colliding sperm whales must absorb the shock to their brains and skulls.[3] He and his colleagues noted scars cover the cetaceans' snouts despite their tough hides.

In spring 2003, Carrier's brother and the pilot Lanham met at a friend's house, when the conversation turned to sperm whales. Carrier learnt of Lanham's story when his sibling alerted him. 'We knew that our ramming hypothesis had some merit and looked into the available technology to test it,' says Carrier.

An interdisciplinary research team, including Carrier and Panagiotopoulou, used structural engineering principles and computer simulations to test how the sperm whale's head might withstand the impacts of ramming. They improvised an engineering method used to assess the endurance of bridges and tunnels.

A series of vertical partitions made of connective tissue in the junk sac may reduce stress to the skull and protect bone and soft tissue during collisions, conclude the researchers. Without this structural support, the 45 per cent greater pressure to the head's bony skeleton could cause fatal fractures.

'Although male sperm whales may not fight frequently,' says Panagiotopoulou, 'we know that aggressive ramming behaviour is a common characteristic in the group of mammals from which whales are derived – the even-toed ungulates, the artiodactyls. A closer look into the anatomy of the heads of other species of whales that ram may reveal a variety of protective mechanisms.'

However, Rendell isn't convinced. In other species, males that battle it out for the favour of females have weapons that are disproportionately large relative to body size. In sperm whales, both sexes have massive snouts, superbly adapted for echolocation.

Panagiotopoulou and Carrier counter that sexually dimorphic traits evolve for a reason and they could have multiple functions. They suggest one reason biologists haven't witnessed a ramming incident between cachalots may be their location. A spectator would have to have a bird's-eye view of the water surface to watch it.

'The notion that current observation methods would have missed a significant amount of head-butting in males does not bear scrutiny,' counters Rendell. For instance, he and his team watched a pair of northern bottlenose whales[4] pound each other with their noses from the vantage of an elevated platform. The bony anatomy of the males' heads can withstand such a battering, but cachalots have no such structure to protect their sensitive navigation tools.

After the publication of Carrier and Panagiotopoulou's study, another team recounted a sperm whale ramming its research vessel, the *M/V Kahana*, near the Mariana Islands in 2007.[5] What's more, it had a video recording to prove it.

A bull peeled away from its pod and sped at least 150 metres to smash into the ship. The craft suffered no damage, and the animal didn't follow-up on its first attack. All three bulls in the group had fresh rake marks over their heads, probably from skirmishing with each other. Whether the colliding whale saw the ship as yet another rival, a threat to a mother and calf in the pod, or a target to misdirect its hormone-charged aggression, there's no doubt it deliberately rammed the vessel.

'It is possible that such contests could escalate to physical ramming in extremis,' says Rendell. The whales would not risk seriously damaging their communication and echolocating organs by battering their noses. 'There would be strong selection to settle things without resorting to physical conflict on a routine basis.'

Maybe Moby Dick pounded the whaler *Pequod* which threatened his life. But he probably didn't ram his head against another male to settle territorial or mate disputes. For that he had teeth.

28

The Giraffe's Neck Solution[1]

THE giraffe's six-metre height and large leopard-like spots make it a distinctive animal of the African savanna. Even as other herbivores crane their necks to reach the lower branches of trees, this towering animal takes advantage of its stature to browse on tree crowns.

Being the world's tallest animal comes at a price. Blood fights gravity to course up the two-metre-long neck to the brain. How does the heart push the fluid so far up? For years, researchers puzzled over this question until they discovered[2] the giraffe's muscular eleven-kilogram heart pumps with such force that the animal's blood pressure is two and half times greater than ours. While this hypertension works as long as the giraffe stands upright, fulfilling an everyday need – drinking water – can turn into a life-threatening situation.

The lanky animal splays its front legs and lowers its head to reach water. The high pressure ought to send blood gushing to the brain, causing a stroke. If it survives this and slakes its thirst, it faces another danger. Standing up again would cause pressure to drop and the giraffe to faint. It has a neat workaround to solve these risks: the jugular vein locks off blood to the head when it bends down. When the head lifts, the vein opens again.[3] Besides, the animal has a network of blood vessels, called rete mirabile,[4] at the base of the skull to regulate blood pressure.

That's not all. As anyone suffering from high blood pressure knows, the disease causes the legs to swell with fluid, a condition called oedema. Yet, the giraffe's long spindly legs show it has solved this problem too. The thick walls of the arteries below the knees endure the pressure, and the animal's hide wraps tightly around it, like compression stockings, giving no space for any fluid to collect.[5]

The primary cause of kidney disease and failure in humans is hypertension. But giraffe kidneys show no signs of damage. The renal capsule, a tough layer of fibrous tissue that envelopes the organs, is strong enough to withstand the high blood pressure.[6]

Despite all these adaptations, giraffes aren't born with hypertension. They develop it as they grow and their necks start to extend.[7]

Besides challenging the cardiovascular and renal systems, the long neck also taxes the skeleton and muscles. Nineteen bones support the necks of flamingos. But the giraffe has the same number of vertebrae as we have: seven.[8] What it lacks in quantity, it makes up in size. Each one is almost a foot long. And yet, the neck is flexible enough that it twists around and rests on the rump like a swan when the giraffe sleeps.

The nuchal ligament, a band of elastic tissue running down the length of the neck, bears the weight of ungulates' heads. In giraffes, it is much larger and more robust, the better to hold up the weight of the head and neck.[9]

The four species and five subspecies of giraffes have all these variations even though they separated from each other two million years ago.

What are the genetic underpinnings of these adaptations? How different is the genome of this unique creature compared to other herbivores? A group of molecular biologists led by Douglas R. Cavener from Penn State University, US, and Morris Agaba from the Nelson Mandela African Institute of Science and Technology, Tanzania, investigated. They sequenced the genes of two Masai giraffes (*Giraffa tippelskirchii*) from Masai Mara, Kenya, and an okapi foetus from the White Oak Conservation Center, US, and scoured them for genetic clues.

The giraffe shares little family resemblance with the okapi (*Okapia johnstoni*), its closest and only surviving relative. The reddish-brown animal with a much shorter neck resembles an antelope with the striped legs of a zebra. Biologists think the neck of their common ancestor may have been longer than an okapi's but shorter than a giraffe's.

Cattle ancestors separated from giraffe-okapi ancestors twenty-eight million years ago. So, the researchers compared the genes of giraffe and okapi with cattle DNA to identify areas of difference. Then by matching the genes of the okapi and the giraffe, the researchers narrowed down the genes specific to giraffes. They estimate giraffes grew tall and evolved a 'turbo-charged' heart within the last 11.5 million years when their lineage split from the okapi's.

The scientists made a surprising discovery. Of the thousands of genes, only seventy make the giraffe the unique animal it is. Several of these regulate skeletal, cardiovascular and neural development, while others affect metabolism and growth. Four genes may govern the development of the spine and legs while eight are likely to administer heart functions.

Mutations in gene MDC1 kill defective cells and protect the animals from cancer. It 'exhibits the most radical evolutionary change in giraffe and okapi compared with all other vertebrates,' write the researchers.

The real surprise was some genes perform diverse functions. One particular gene, FGFRL1, regulates the growth of the skeleton but also influences the cardiovascular system. The ones that control digestion of the nutritious but toxic leaves of the acacia also affect the heart.

Biologists debated whether giraffes grew tall first and then evolved the heart adaptations to counter the difficulties posed by the long necks. Since many genes affect more than one function, the researchers suggest the changes didn't occur independent of each other but in tandem.

While genes and amino acids show how giraffes got their long necks, further studies have to confirm their functions.

'We have opened a can of genes,' says Agaba, 'and I believe that something good and useful will come out that may advance our knowledge of nature, our health and that of our ecosystems on a fragile earth.'

The team writes that studying the giraffe's adaptations would provide insights for the treatment of cardiovascular disease and hypertension in humans. One hopes that growing a long neck isn't a precondition for the cure.

29

Airborne Sleep[1]

GREAT frigate birds (*Fregata minor*) wheel around in the skies for days, riding thermals, and keeping a sharp eye out for prey in the open oceans. Built like ultralight planes, the metre-long birds weigh less than 1.5 kilogram and have a wingspan of over two metres. The birds' ratio of wing reach to body weight is greater than that of any other bird. Their lengthy, narrow, pointed wings are ideal for soaring high, and they seem to prefer staying aloft even when land is in sight.[2]

Flying confers an advantage on birds. They don't have to climb trees and mountains or wade across rivers and oceans. Flapping to create lift, they fly over barriers that frustrate land and sea animals. But being airborne is an energy guzzler. Migration burns up reserves even though the flyers feed at regular pit stops along the way. Some lose a quarter to half their body weight in making these distant journeys.

How long do frigate birds remain aloft? How do they power their aerial lifestyle?

A team led by Henri Weimerskirch, an ecologist at Centre d'Etudes Biologiques de Chizé, France, tracked forty-nine frigate birds with solar-powered satellite transmitters, heart-rate monitors and accelerometers for four years.[3]

Between June and September, birds without parental duties left Europa Island, between mainland Africa and Madagascar. They caught the trade winds north to the equator and banked eastward towards Indonesia. Scientists know frigate birds ride the thermals. However, the skies above balmy equatorial waters are still. Called the doldrums, they were the bane of pre-Industrial era sailors. The high-flyers skirted them by riding the trade winds sweeping from the southern and northern hemispheres.

Rising columns of warm humid air condense to form cumulus clouds. The birds caught this current and spiralled upwards without an effort to the base of the clouds at 700 metres. While they coasted horizontally from these heights, they frequently continued their corkscrew flight pattern inside these clouds to reach more than 4,000 metres.

Any activity at high altitudes is laboured because of low oxygen levels. Since the birds let the air currents do their work for them, they didn't spend much energy and their heartbeats didn't speed up.

One young bird stayed airborne for two months. Most coasted non-stop for up to forty-eight days, travelling an average of 420 kilometres a day. They rested for eight to forty-eight hours in the islands of Chagos that lie south of Maldives, Indonesia or the Seychelles on their great transoceanic flights.

During the day, the frigate birds fly above the sea surface to catch fish. However, they suffer from a critical design fault – their feathers are not waterproof. Should they land on water, they get waterlogged and cannot take off. Neither can they dive after fish. By necessity, they catch prey like flying fish on the surface and gulp it down midair. Although these feeding bouts consume only about 10 per cent of their time, they drain energy because the birds flap their wings to gain altitude before diving through the air to catch the next fish.

Such aerial prowess takes time to master. Parents spend up to two years raising their chicks. This is far longer than any other bird.

If frigate birds spend months on the wing, when do they sleep? They cannot touch down on the water like albatrosses. Their body design leaves them with no choice but to keep flying when they are far from land. Sleep is necessary to restore physiological functioning of the brain. Do they doze in flight?

Another group of scientists led by Niels C. Rattenborg, of the Max Planck Institute for Ornithology, Germany, monitored brain wave activity by implanting custom-made electroencephalograph sensors (EEG) and accelerometers on fifteen nesting female great frigate birds in Genovesa Island, Galapagos. After five to ten days, when the birds returned to the nest after foraging at sea, the researchers retrieved data from their instruments.

The scientists discovered that an hour before sunset, the birds gained peak altitude to compensate for the steady downward drift through the night. Although their flight pattern of circling up thermals and coasting down offered opportunities to snooze, they didn't take them. The fear of colliding into others or crash landing in water appeared to keep them wide awake.

When they did doze, they preferred to have one side of their brains active and the opposite eye open to watch where they were going. But sometimes they let both hemispheres rest and switched off totally. Their snooze, like that of mammals', was in two phases: slow-wave sleep, also called deep sleep, and the brief random-eye-movement (REM) sleep.

On the upward spiral, half of the brain and the corresponding eye dozed for twelve seconds at a time. When the frigate birds glided down, losing altitude, they napped for seven seconds only. The longest any of the birds slept in a bout was a little over five minutes. Such fleeting episodes of shut-eye totalled forty-five minutes a night, accounting for 3 per cent of their time spent in the air.

Within those few moments of sleep, the birds went into five-second bursts of muscle-relaxing REM sleep. But this deactivated both halves of the brain. The sleeping birds didn't plummet but stayed aloft with wings outstretched, flying on autopilot.

On average, frigate birds slept fewer than eight hours over ten days in the air. Despite fragmented naps for many consecutive nights, they suffered none of the consequences such as drowsiness and inattention. Once at the nest, they caught up by snoozing for over 50 per cent of the time.

The researchers speculate that the half-awake birds may arrive at good fishing spots by daybreak. Or they may even fish on full moon nights.[4] Under pressure to return to their nests with enough food to feed a chick and relieve their starving mates, they cannot afford to nap.

Once the offspring at Europa Island become independent, they leave their island home and circle over the Indian Ocean, staying airborne for months. One youngster flew large sweeps over the Indian Ocean, clocking 55,000 kilometres in six months. In all that time, it rested on islets for only four days.

Frigate birds challenge the long-held belief that a full night's sleep is essential for the brain to function normally. They feel no compulsion to land when they aren't breeding. The young bird that remained aloft for two months may have snoozed longer on the wing.

More than their skill at catching a few winks while airborne, the great frigate birds' ability to get by on power naps alone is incredible.

30
Learning from Monkeys?[1]

BEARDED capuchin monkeys (*Sapajus libidinosus*) flock to wild, fruit-laden cashew trees in Brazil. Fibrous shells covering the kernels exude a caustic resin that causes blisters. Humans char them before eating the meat. Since capuchins haven't figured out the use of fire, they developed different ways of dealing with the corrosive nuts.[2]

At Fazenda Boa Vista in north-eastern Brazil, the monkeys rub the fresh spongy shells on rough-barked trees to wear them down, exposing the kernel. They dig out the meat with their index fingers. With dry hard ones, they opt to hammer them open with rocks.[3] Researchers think this is the most difficult tool-use to master even for apes.

Humans thought for a long time that they were the only ones to wield implements. That view changed dramatically in

1960, when Jane Goodall observed chimpanzees fishing ants with grass stems. Since then we have learned that a whole range of animals from crows to dolphins and various primates utilize objects to get their way. Now capuchins are the latest to join these ranks of tool-users.

The capuchins don't smash with all their strength. Instead, they pound with moderate force and by gauging the condition of the nut, strike the next blow. To keep the kidney-shaped drupes from rolling, the primates place them just right and employ large flat stones as anvils.[4] Some adults set the nuts with the concave-side facing down and clobber the arched end.[5] Although this skill takes time to acquire, the shell is easier to crack.

About kilometres km away as the crow flies, the capuchins of Serra da Capivara National Park use rocks to break open the shells of fresh and dry nuts. After feasting on the kernels, they dump their tools at the site. When the next fruiting season arrives with the summer and the monkeys need nutcrackers, they pick their favourite stones from these collections.

Researchers from the UK and Brazil led by primate archaeologist Michael Haslam investigated how long bearded capuchins had used tools.

'We know so much about the technological evolution of humans but so little about the evolutionary history of other primates,' says Lydia Luncz of the University of Oxford, UK, one of the researchers. 'To find answers to the question – why are we humans so technologically advanced – we need to compare the evolution of our own behaviour with the evolution of other primates. We use stone tools as a window into past behaviour of wild primates.'

In May 2013, the experts mapped four locations of discarded rock implements within Serra da Capivara National

Park. The objects were of a particular size and kind: smooth, hard quartzite for hammers and flat sandstones for anvils.

The scientists weighed these large, stained and worn nut-smashers. Anvils were approximately 600 grams on an average, while hammers were 200 grams. But the stones strewn around the area were only sixty grams heavy. From where did the capuchins get their tools? They hauled them from a nearby dry stream bed up to the base of cashew trees.

The researchers then excavated one of the four sites with trowels and shovels. They uncovered sixty-nine stone implements within a 35 square metre hole that was 70 centimetres deep. Using gas chromatography-mass spectrometry, they confirmed the dark stains on them were residues of cashew resin such as cardanol 3 and cardanol 4.

Charcoal, the result of natural fires, found with the rocks was ideal for carbon-dating. The researchers gauged the tools belonged to three distinct ages. Modern ones littered the surface. Underneath that layer were stones from as far back as the seventeenth century. Rocks dating from the thirteenth to fifteenth centuries lay buried at the bottom. The oldest implements showed a history of 600 to 700 years of tool use by capuchins, making them the most ancient wielded by animals outside Africa. The monkeys were pounding these nuts long before the European colonization of South America.

Were the cashew trees so old that generations of the primates returned to the same spot?

'Individual cashew trees don't last that long,' replies Tiago Falótico of the University of São Paulo, Brazil. 'We know from pollen evidence that cashew trees were present in that valley for at least 7,000 years. We assume that places with clusters of cashews in the present would be likely to have had cashews in the past too.'

Generations of capuchins didn't change the basic contours of their tools. The oldest implements were of similar weight and material as modern ones. Just like chimpanzees (See Chapter 7: Conservative Chimpanzees), these primates are conservative, reluctant to make any changes.

The team also dug at Fazenda Boa Vista, where the capuchins' preferred method was rubbing cashew nuts on rough surfaces, but they found few artefacts. 'We believe that it has to do with the availability of suitable stone hammers,' says Luncz. 'The capuchins there crack open hard oil palm nuts which require large tools to open them. Occasionally, they also crack open dry cashew nuts. Those tools are very rare in the landscape. Therefore, they are very valuable, get re-used intensively and never enter the archaeological record.'

Another primate with a long tradition of using implements is the chimpanzee. Archaeologists excavated 4,300-year-old stones utilized to break open palm nuts from sites in Côte d'Ivoire.[6] Scientists then wondered who invented the skill. Was it a human ancestor or a common progenitor of humans and chimpanzees? Or did ancient chimpanzees watch humans and imitate them? Or, perhaps, such technology evolved independently in primates and humans.

In Africa, chimpanzees entering cultivated cashew plantations eat the cashew apple but discard the nuts. Only humans and bearded capuchins have figured out how to tackle the hard shells.

In Asia, long-tailed macaques of Thailand use stones to crack open oysters, marine snails, crabs and nuts. In another study, Haslam and his team say these primates also choose their tools carefully, sometimes using the same rock to work on several molluscs.[7] However, these rocks aren't antiques; they are less than fifty years old.

The capuchins in Serra da Capivara are dexterous with implements in other contexts. They use stones to dig for tubers and spiders,[8] bang them against rocky outcrops to intimidate a researcher,[9] and some females throw them at males as a sexual display.[10] One used a grass stem to pick her teeth and nostrils.[11]

The long history of capuchins' usage of stone hammers poses more questions about the origins of tool use and its spread in the New World. Perhaps, the first humans, who arrived in South America almost 15,000 years ago, learned cashew nuts were edible by watching the primates.

People living around the national park ate cashews at least 7,000 years ago. To prove whether humans learned to open cashew nuts from monkeys, the team would have to find capuchin tools of an even older provenance.

If it did discover such ancient nut-smashers, it would make a case for 'human see, human do'.

31
Honey-Collecting with Birds[1]

A small, brown bird lands on a low branch beside a hamlet of Yao hunter-gatherers in Mozambique. Its loud insistent two-note trilling call, flapping wings and flashes of prominent white spots on its tail attract the attention of a few men, who set out after it. As its name suggests, the myna-sized greater honeyguide (*Indicator indicator*) leads them to beehives.

Bees typically hide their nests high inside hollows of large trees like baobabs and in sheltered rock crevices. Although honey can make up to 10 per cent of indigenous people's diets, finding hives in the deciduous forests and savannas of Africa is time-consuming. Having a bird locate and then direct people to them makes the process so much easier.

The honeyguide lives on beeswax and the brood of honeybees. It appears to have some defence against bee stings such as unusually thick skin and upraised nostrils, but it cannot tackle enraged bees by itself. It depends on humans to scare the insects away with smoke and open up the nest with axes.

Leading the Yao men, the honeyguide flits from tree to tree toward a beehive as a pet dog would lead its human to the cupboard with its favourite treat. It often flies sorties to the nest and back. Throughout the walk, the men 'chat' with the honeyguide by making a loud trill and grunt, 'brrrr-hm'.

Once the party nears the destination, the bird perches close by quietly or sometimes makes another call as if to say, 'We've arrived'. Sometimes, it circles the tree to show the location.[2] Until the men harvest the hive and leave, a process that may take as long as ninety minutes, it waits patiently before feeding on the fat-rich wax combs left behind.

The birds' call and display urge hunters to follow them throughout their range in Sub-Saharan Africa.

This unique collaboration between a free-living creature and local communities was known for centuries. In 1609, João dos Santos, a Portuguese missionary, published an account of his travels through south-east Africa in *Ethiopia Oriental*. He wrote of a small bird entering his mission church to nibble on candlesticks and also lead people to beehives.

Evolutionary biologist Claire Spottiswoode of the University of Cambridge, UK, and her colleagues worked with the indigenous Yao men to investigate whether honeyguides were reliable.

As the honeyguides led the hunters through the vast Denmark-sized deciduous woodland called Niassa National Reserve in Mozambique, the researchers trailed along, marking their progress with a GPS logger. Since they covered a large area and individual birds had home ranges not exceeding one square kilometre, the authors suggest a different bird guided them every time.

The birds led them to at least one beehive in 75 per cent of ninety-seven guiding events. Most hives belonged to the common honeybee and the rest to stingless bees.

Spottiswoode and her colleagues uncovered another astonishing layer to this mutually beneficial relationship. They conducted seventy-two trials with three different audio recordings: song of the ring-necked dove, normal Yao words, and the 'brrrr-hm' call. The birds showed no interest in the cooing of doves or human vocabulary but recognized the trill-grunt as an invitation to hunt. Since the men don't make this sound on any other occasion, the birds appear to know it well. The probability of finding a beehive more than tripled if the men chatted with the birds.

'Honey hunters use special calls to signal to honeyguides that they're eager to follow,' says Spottiswoode, 'and honeyguides use this information to choose partners who are likely to be good collaborators. So, it is a two-way conversation between our own species and a wild animal.'

Other animals like common bottlenose dolphins and Irrawaddy dolphins collaborate with fisherfolk by driving fish towards fishing nets. But the greater honeyguide's relationship with humans is described as 'possibly the most advanced bird-mammal relationship in the world.'[3] By using their distinct calls, birds and men communicate their intent to each other.

Honeyguides don't have the option of learning how to collaborate with humans from their parents as the honey-hunting men do. Like cuckoos, the birds lay their eggs in the nests of other species, and their foster parents raise them on a diet of insects. So, the opportunity to learn from their biological parents doesn't arise.

Besides, people of different regions summon the birds with different calls. A honeyguide from Mozambique cannot recognize the melodious calls of the Hadza people in Tanzania. The Boran pastoralists of northern Kenya call the birds by whistling through closed fists, snail shells and hollow palm nuts. To adapt to the varying human calls, the birds can't rely on an innate knowledge of a specific man-made sound.

How do the birds learn the special calls of honey hunters of their area? The authors suggest the birds may learn by watching other honeyguides near beehives.

'We don't know whether this behavioural specialization reflects a honeyguide 'culture',' adds Spottiswoode. 'This is just a hypothesis. In fact, culture has a very precise biological definition that would take careful experimental work to test in honeyguides.'

American and British researchers speculate that the collaboration may have arisen in the Pliocene, about three million years ago, between honeyguides and one of our earliest ancestors, Ardipithecus or Australopithicine.[4] The birds developed this specific repertoire of displays and calls after identifying humans as their best partners in this endeavour. They also solicit ratels,[5] but these incidents seem to be uncommon.

This millennia-old tradition may be petering out as people switch to store-bought sugar and honey. When no one shows interest in their calls, honeyguides stop soliciting people to go honey collecting. The birds had already given up guiding three decades ago in some parts of Kenya. Spottiswoode says the tradition is on the decline in Zambia too. As people are excluded from using national parks, these relationships may cease altogether. With these disappearing relationships, the opportunities for understanding them are also declining, bemoans Spottiswoode. 'It mostly persists and thrives in wildernesses like Mozambique's amazing Niassa National Reserve where people and wildlife still coexist.'

Without human partners, honeyguides may not have access to beehives. How do they survive?

'We don't have data on this,' replies Spottiswoode, 'but honeyguides certainly do eat insects too.' African bees commonly 'abscond', moving nests to a new site. If the abandoned hive is in an accessible location, such as a tree hole with a big entrance, then the birds might be able to reach the wax combs without any human help.

This is a rare instance of a free-living creature with no prior training cooperating with humans to harvest a natural resource.

32

Are Humpback Whales Altruistic?[1]

A pod of killer whales, or orcas, charged in formation at a lone Weddell seal resting on an ice floe. The seal may have been scared, but it was above water and out of reach of the predators. The phalanx of whales pushed a bow wave ahead of them as their thumping tails created another swell. When they approached the ice, the waves crashed over it and the seal slid into the Antarctic waters. It was as good as dead, but it made a desperate break for a nearby pair of humpback whales (*Megaptera novaeangliae*).

One of these behemoths rolled over on its back and caught the seal on its chest. Intent on snatching their prey, the orcas gave chase. The whale rose out of the water, pushing the seal into the air. But the hapless little animal slid down the slope of the gigantic body. The humpback shoved it to the safety of its belly with an enormous flipper. What could the pursuers, half the length of humpbacks, do? The stand-off continued for several minutes before the frustrated predators left. The seal made its way unharmed to a large ice floe.

The humpback's solicitous behaviour intrigued a human observer. Marine biologist Robert Pitman of the National Marine Fisheries Service, US, had already witnessed two such incidents.

During the preceding week, he had seen a pair of humpbacks prevent killer whales from making a meal of two seals off the southern tip of South America. Best known for their haunting songs, the humpbacks warded off the predators by bellowing and slapping the water with their flippers and tails.

Killer whales don't prey on seals alone; they also fall upon the calves of humpbacks and other large whales. Protective humpback mothers holler and charge at their tormentors while slapping their tails and flippers on the water surface. They hoist their calves into the air and perch them on their heads or backs. The same behaviour also rescued the seals. Outnumbered, lone humpback mothers cannot always save their young.

In the North Pacific, orcas kill 18 per cent of humpback whale calves less than a year old. In some populations, up to 40 per cent of the fifteen-metre giants had scars sustained during their calf-hood from the predators' attempts on their lives.[2] The coordinated hunts of the seven-metre killer whales earned them the sobriquet 'wolves of the sea'. To safeguard their young, humpback mothers take circuitous routes through areas avoided by killer whales.[3]

When they come under attack, the large slow-swimming whales cannot outrun their pursuers. They can't fight back with menacing teeth because they have none. Instead, they have closely spaced vertical baleen plates to sieve shrimp-like krill from the water. Their enormous five-metre-long flippers, a third of their body length and weighing a ton, are the largest among marine mammals. Barnacles encrusted on the leading edge of these limbs can gouge out the flesh of their opponents. The gigantic crustacean-studded flippers in the front and the powerful tail fluke at the rear are the humpbacks' best weapons.

'These ponderous appendages are potentially a lethal threat, and killer whales keep their distance around excited humpbacks,' says Pitman.

Pitman wondered if humpbacks commonly save other species. He and thirteen colleagues compiled 115 cases of humpbacks confronting killer whales documented by at least fifty-four observers between 1951 and 2012 in the Pacific and Atlantic Oceans.

Humpback whales usually go out of their way to rescue their own kind. On one occasion, three swam seven kilometres to converge on a pod of killer whales. They couldn't have seen their foes from that distance, but they may have heard the orca calls to coordinate an attack or summon reinforcements, say the researchers.

However, humpbacks don't rush to just any killer whale call. Orcas are predominantly mammal or fish eaters and the calls of these two groups are distinct. Humpbacks seem to know this and surge toward meat-eating orcas, harassing them and sabotaging their hunts. When they meet the pescatarians, they get along fine.

Since they charge to the scene on hearing long-distance sounds, they can't identify the victim species. If the prey animals were not humpbacks, most persisted in chasing away the predators. Almost 90 per cent of such rescues were of other species, including other whales, seals, sea lions and an ocean sunfish. One male humpback, recognized by unique marks on its tail fluke, rescued Stellar sea lions on two occasions separated by fifteen years.

Humpbacks rushing to the aid of others often work in pairs. In one spectacular case, thirteen of them ganged up to protect a humpback calf. They enclose the calves within a ring by pointing their heads in and tails out. If they are alone, they think nothing of taking on killer whales ten or more strong.

Each instance of help may last an hour, sometimes as long as seven hours. They forgo opportunities to feed, rest and socialize.

The humpbacks were not always successful in saving others' lives. In a few cases, they arrived too late to prevent the preys' deaths.

There are two reasons why researchers have only now realized the humpbacks' motives.

'Whales are still recovering from near extinction due to twentieth century whaling,' says Pitman. Few people are out in the oceans observing them. When they do see these interactions, they suffer from observer bias, drawing conclusions based on their expectations. 'When people, even researchers, see spirited interactions between humpbacks and killer whales, they usually assume that the killer whales are attacking.'

Humpbacks don't target killer whales alone, they also mob other predators like false killer whales and pilot whales. But not all humpbacks are the first responders of the ocean. Many hang back as bystanders. Sometimes, the ones that wade into the fray ought to be keeping their distance. On two occasions, mother humpbacks with calves joined others to chase away killer whales from sea lions. Why did they put their calves at risk?

Pitman suggests the ones charging into these confrontations are individuals that experienced skirmishes with orcas when they were young. 'My hope is that with this new perspective on humpback–killer whale interactions, we will have answers to some of these questions before too long.'

Are humpbacks being altruistic?

Humpbacks may mob a predator, much like birds chase raptors. Although they live solitary lives most of the year, they may bond with others of their species when they congregate at feeding grounds. Going to the rescue of others could build reciprocal relationships. If one whale helps another, others may come to its rescue when the time comes. But it is unlikely that any of the other species they save from death would return the favour at a later date.

What makes them assist creatures other than humpbacks at considerable risk and energy cost? The authors speculate it may be a spillover of the same behaviour, but this needs more research.

If orcas are the wolves of the sea, humpbacks have earned the sobriquet 'Good Samaritans of the ocean'.

33

The Playful Keas of New Zealand[1]

FOR keas (*Nestor notabilis*), a species of large parrot, life is a game. In the highlands of South Island in New Zealand, they jostle, kick and bite their companions, sometimes dragging them across the ground. They roll on their backs, their feet up, while their partners jump on their tummies. After locking their long, hooked bills, the olive green parrots twist and push, performing a bird version of arm wrestling.[2] They fashion toys out of inanimate objects, playing tug of war or flipping rocks and sticks to each other. For their comical feats, they earned the epithet, 'clever clowns of the mountains'.

The urge to gambol is ingrained, and if one partner attempts to stop, the other protests. Throughout these fun and games, whether in the company of others or alone, they warble. It is a distinctive sound called 'play call' reserved only for recreation.

Is it an invitation to a game?

Dogs, for instance, do the play-bow, crouching low with their snouts touching the ground and rear ends held high, to provoke another to frolic. Keas solicit play too. They cock their heads to one side while hopping sideways towards their partners. They then roll on their backs, enticing their mates to pounce. Could the warble be the audio counterpart of this invitation?

Raoul Schwing, of the University of Vienna, Austria, and colleagues from Australia and New Zealand studied the call's role in kea games at Arthur's Pass National Park. Schwing chose birds that weren't already fooling around as his subjects. He played recordings of different sounds – the play call, a kea screech and whistle, the song of the unrelated South Island robin and a standard monotone – through a speaker for five minutes and watched their reactions.

Except for the warble, the parrots reacted to none of the other sounds. When they heard the play call, they didn't seek to join any games in progress. The researchers concluded it doesn't mean 'Come, let's play'. Instead, the parrots grew frisky with the ones closest to them. If they were alone, they turned any object they picked up into a play accessory.

'When I saw a kea spontaneously pick up a rock and start playing with it, I knew we had found something special,' says Schwing.

If nothing caught their fancy, they performed aerial acrobatics.

The effect of the warble is akin to yawning in humans. One person opening their mouth wide triggers others to repeat the act. But the keas' play call elicits much more than imitation. When humans hear laughter, their moods lighten. The parrots' vocalization has the same impact on others. This species is the first non-mammal, to scientists' knowledge, capable of changing one another's mental state.

The researchers expected the play call might be a sporty invitation. They didn't expect to find it infectious.

Why did this distinct cry evolve?

Unlike humans and bonobos, chimpanzees, macaques and rats don't wrinkle their faces when they romp. With no facial expressions to signal their friendly intentions, they perhaps evolved specific sounds. The keas' playful warble may have a similar purpose.

In most creatures, no-nonsense adults have little time for frivolous horseplay. Adult male and female bonobos play during courtship. African hunting dogs and timber wolves frolic to strengthen their packs' social bonds. Otherwise, full-grown animals don't romp with members of their own or the opposite sex. They have more serious business to attend to such as guarding against danger, scrounging for food and watching out for rivals.

Adult keas, however, need no excuse to play. Aren't they under the same compulsions as other animals? They relax their guard because they fear nothing. 'Their main predator might have been a now-extinct Haast eagle with a wingspan of about three metres,' says Ximena Nelson, a collaborator in the study. With zero threats to look out for, they can afford to indulge in 'leisure time', especially in spring when food is plentiful.

Full-grown keas pair for life, but they congregate at feeding areas. Misunderstanding and aggression could be the bane of such gatherings. But keas' social life is one of peace and goodwill. Maybe frolicking builds bonds between individuals. Schwing says he hasn't seen an altercation break out between these wild parrots. In a flock, appropriately called a circus, the warble broadcasts the caller's playful intentions and smoothens group interactions.

On hearing Schwing's play call recording, keas gambolled only for a few seconds. For a bird that seems to live for revelry, these bouts were too short.

'I think that the drop of play was based on the generally unfavourable weather conditions,' explains Schwing. 'If we had conducted the experiment in summer, the effects might have been much longer lasting, as the natural propensity to play would have been higher.'

Conducting experiments with a species as smart and curious as keas can be aggravating. They were intent on destroying the researchers' audio playback equipment. Nelson says they 'kea proofed' the speakers with thick wire.

Perhaps they thought the gadgets making the play call wanted to have fun.

34

Peeing on Eggs[1]

THE nightlife of New York City prepared frog expert Jesse Delia for studying the mysteries of amphibian parenthood in Central and South America. He used to go clubbing until sunrise and take the bus to school or work. 'We call this "breaking night" where I'm from.' Years of partying helped him stay awake until the wee hours to study frog behaviour.

Delia, from Boston University, US, and Laura Bravo-Valencia, of Universidad de los Andes, Colombia, observed forty species of glassfrogs at twenty-two sites across Mexico, Panama, Colombia, Ecuador and Peru. Over six rainy seasons, they jogged up and down slippery rocks in cascades along streams in hot lowland forests and cold mountains in the Andes.

Glassfrogs, named for their transparent bellies that reveal their internal organs, weren't thought to dote on their offspring. Less than 20 per cent of frogs show any degree of parental care.

The 150 species of glassfrogs glue their eggs to shrubs hanging over rivers and streams, convenient for hatching tadpoles to drop into the water. Their parental duties ended with finding the right spot for the next generation, thought biologists. If any species showed concern for their young, it was only males such as Lynch's glassfrogs (*Centrolene lynchi*).[2]

Female glassfrogs absorb dew from leaves through the belly skin until their bladders fill up. In amplexus, or frog sex, males ride piggyback on these bloated females, fertilizing the eggs their mates spew. Once the job's done, the mothers stay to urinate on the mass of pearly white globules. Hours later, the eggs balloon four times their original size and develop a thick coat of jelly that insulates them from desiccation during dry weather.

The researchers discovered that every species cares for its progeny. In most cases, mothers brood the eggs for one night. Over the following days, rainwater keeps the gelatinous layer moist. The developing embryos of some species need long-term tending as the thin outer coating does not retain water. Their fathers become single parents, urinating on them on alternate nights for weeks. In only one species, Magdalena giant glassfrog (*Ikakogi tayrona*), does the mother provide extended care.

Females of some glassfrogs (genus *Centrolene*) stay for some time with their eggs but apparently don't hydrate them. It's the males that transport water. The researchers speculate the female behaviour of tending eggs may have all but disappeared in these species after their mates took over parental duties.

Delia explains co-evolution may be the reason for these different parenting styles. 'As parents provide more than one day of care, offspring become dependent. Thus, the two could co-evolve leading to really needy eggs and parents that provide long care durations.'

Grainy Cochran frog (*Cochranella granulosa*) and Chiriqui glassfrog (*Teratohyla pulverata*) mothers spend only about two hours with their eggs. Although brief, brooding behaviour is critical, say the scientists.

When the researchers touched females with no responsibilities, the frogs fled. But mothers with eggs sat tight even as they were pushed. One grainy Cochran frog returned every time they shoved her off. When the researchers left them alone, the devoted mothers came back to their eggs.

How important is parental peeing for developing embryos?

The biologists removed twenty-four mothers of each species soon after they laid their eggs, without giving them a chance to urinate on them. They observed the same number of other females that continued to stay with their brood and compared the outcome for the progeny. The abandoned eggs became dehydrated or infected with fungus.

The un-urinated frog eggs had thin coats. The researchers fed them to captive katydids that had no trouble munching them down. However, the insects couldn't get through the jelly coating of the embryos nurtured by their mothers.

Even if eggs survived abandonment, it affected them in other ways. Freshly hatched tadpoles nurtured by their fathers are ready to begin eating. In 2014, Delia found that tadpoles of northern glassfrog (*Hyalinobatrachium fleischmanni*) hatch early when they detect their fathers had ditched them.[3] Smaller than average with undeveloped digestive systems and unable to feed with their mouths, they rely on yolk reserves to survive. This may lower the odds of their survival.

To unravel this behaviour, the researchers had to muck about all night for months and get drenched in the rains. How did they cope?

Fieldwork became meditative, Delia says, as his thoughts centred on glassfrogs and estimating where and when he had to be on certain sections of a stream. 'Humans are strange,' says Delia. 'We seem to get personal reward from the most miserable tasks.'

The trigger for this study was a chance discovery six years ago while Delia was out frog-watching all night.

'I had been monitoring eggs of *Cochranella granulosa* [grainy Cochran frog] every night for a couple months in 2011, and never noticed any parental behaviour,' says Delia. When he waded down a stream around 5 a.m., he saw a female brooding a clutch of freshly laid eggs. This observation challenged the accepted wisdom that this species and its females didn't care for the offspring.

Similar is the case with Lynch's glassfrog. Other researchers reported only males care for eggs, but in this study, the biologists observed that females were the primary caregivers. Why did the earlier biologists get it wrong?

'It's normal to see males really close to eggs,' says Delia. But, he concedes, perhaps they do a little bit of parenting that he and Bravo-Valencia didn't observe.

Every night the researchers made detailed observations throughout the incubation period. Female Lynch's glassfrogs stayed with their eggs between midnight and 5 a.m., a window of time that other scientists may have missed.

Still, all night benders in New York and studying frogs in the wilds of Central and South America couldn't have been more different.

35

No Ants Left Behind[1]

As soon as the ant scout returns after identifying the target, 200 to 500 warriors march out in an orderly column. Once they reach the termite nest, the larger ones, called majors, rip up the concrete-hard mud fortifications, and minors pour into the vulnerable colony, seizing and killing the inhabitants. Soldiers of the besieged nest fight back, clinging to or ripping the legs off their assailants. The invaders are among the largest of the world's ants – Matabele ants (*Megaponera analis*).

The two-centimetre ants, named after the fierce Matabele tribe of southern Africa, suffer high casualties during the hour-long battle. As they head back to their encampment with their dead booty, many of their ranks lie wounded at the site. They are far from abandoned as their healthy comrades, the majors, hoist them away from the battlefield. Such rescue of the injured is unique to social insects.

Why do the Matabele ants bother helping individuals? With a colony size between 900 and 2,300 members, there are plenty of warriors.

Scientists led by Erik Frank, from the University of Würzburg, Germany, investigated how these insects sense others' predicament and why they rescue them. They observed fifty-two colonies of Matabele ants conduct 420 raids in the Comoé National Park, Côte d'Ivoire, for thirteen months over three years.

They measured how fast the maimed insects could run after losing one or two limbs. The amputees couldn't keep up with their healthy comrades. Instead, they stumbled along on their phantom legs. The ones with termites clinging to them were even slower than the disabled.

When the researchers forced the injured to return alone and unaided to their nests, 32 per cent of them died compared to the normal 10 per cent mortality of healthy ants. Spiders, lying in wait along the route, preyed on the wounded. Exhaustion killed the others, while the predatory African stink ant (*Paltothyreus tarsatus*) picked off a few. By being carried, the survival of the injured improved.

Ants communicate primarily through smelly chemical compounds called pheromones. These substances are produced by glands in their rear ends such as poison and Dufour's as well as the mandibular gland near their mouth parts. Which one of these organs sends out a chemical SOS?

The researchers extracted secretions from the poison, Dufour's, and the mandibular glands. They smeared these on dead and live ants and watched how the rest behaved.

The rescuers reacted vehemently to poison and Dufour's gland secretions. They shoved any ant that smelled of these chemicals off the column, perhaps not recognizing them as their own.

However, they went to the aid of living and dead ants covered in mandibular compounds. They dropped the deceased ones after struggling to pick them up. 'A dead ant has all of its legs sticking out in all directions which makes carrying it a lot more difficult,' says Frank. 'We believe that is why they dropped them.'

The 'paramedics' abandoned the healthy for different reasons. Since they weren't hurt, the able-bodied struggled and rebuffed help. A few, however, let themselves be carried.

But the majors lost no time in hauling away the injured smeared with oral secretions. The latter tucked their limbs and curled up, making it easier for their mates to medevac them. So, this substance, the ant equivalent of calling the 108 ambulance service, triggered the helpful behaviour in healthy ants.

Using gas chromatography-mass spectrometry, the researchers identified the SOS compounds as the foul-smelling dimethyl disulfide and dimethyl trisulfide.

Although the ants have plenty of majors and minors to raid termite nests twice, sometimes four times, a day, they minimize their losses by rescuing compatriots. Up to fifteen individuals a day suffer injuries while the nest produces only thirteen young in that time. If they didn't pick up the injured, the colony would shrink in size.

'It would basically have to increase the birth rate by that number,' says Frank.

Having to carry the injured back doesn't place an extra burden on the troops. Only 30 per cent of the raiders carry prey back to the nest, so the rest engage in relief. Since most of the rescued survive, the benefits of helping them outweigh the effort. Wounded veterans form 21 per cent of a typical invading column. The researchers estimate Matabele ant colonies are 29 per cent larger than others that don't rescue their mates.

At the end of a raid, when the minors leave the termite gallery, the majors have a last look around and evacuate the injured. However, some fuss over slight injuries and want a free ride back. They stumble towards their nest, exaggerating their disabilities.[2] At the first sight of their comrades, they collapse and pretend to need help. When everyone ignores them, they man up and race after their mates. The warriors hoist even healthy ants if they are caught dawdling behind the lines instead of marching home.

However, they don't carry every injured one to the safety of the nest. When Frank mortally wounded some by disabling five of their legs, the other insects left them to die. In ant triage, it is not the rescuers but the fallen that decide whether they need to be saved. Perhaps realizing the extent of their injuries, they figure help is futile. They thrash about and ignore their mates, making medevac impossible.

When the raiding party returns, the colony turns into an emergency room, with healthy individuals treating the injured. They pry clinging termites off. The 'nurses' brace the hurt limbs with their front legs while grooming and licking the injuries for up to four minutes. Without these ministrations, 80 per cent die within twenty-four hours. Even an hour of tending makes a marked improvement to their survival, with only 10 per cent dying.

With their mobility restored, the insects resume taking part in excursions out of the nest. Although the ones with amputated legs are handicapped for life, after a day's recuperation, they run almost as fast as healthy ants.

No other non-human animal takes care of its wounded veterans as these fierce insects do.

36

Feigning Death to Avoid Sex?[1]

THE dragonfly squeezed out the last of her eggs in a sheltered thicket of reeds on the banks of a pond and took to the air. A patrolling male with bright blue-and-yellow-spotted abdomen swooped on her. She could make a dash for it, but her chances of outmanoeuvring him were low. She fell out of the sky, crash-landing into a bush, where she lay inert on her back. Seeing her lifeless body, the suitor abandoned his pursuit and flew away to await another. When the coast was clear, the female common hawker (*Aeshna juncea*) lifted off. Not only had she fooled the male but a human observer too.

Rassim Khelifa, of the University of Zurich, Switzerland, was looking for dragonfly eggs at a pond near Arosa in the Swiss Alps in the summer of 2015. The seven-centimetre insects don't normally lie with their bellies up. He went close to one inert hawker to see what was wrong with her, but she flipped over on her feet and flew away. None had reported this behaviour among dragonflies before.

Playing dead is a common defence strategy deployed by many animals, from water bugs to some opossums, to escape their predators. Some go to great lengths to put on an act – contorting their bodies, flipping over on their backs, and hanging their tongues out of gaping mouths. Set right a hognose snake that has turned belly up, and it rolls over on its back again. And for good reason.

Pretending to be lifeless is the last resort of desperate prey, even if it means allowing a curious predator close enough to chomp. Movement triggers predatory behaviour. A stalking cat is most likely to ignore a quail faking death and continue stalking one that's running. A pygmy grasshopper (*Criotettix japonicus*) erects prominent spines while pretending to be dead, a painful morsel for a frog.[2]

While such pretence to escape predators is common, using the same strategy to avoid sexual encounters is rare and restricted to a handful of insects and spiders.

Fearful male spiders offer silk-wrapped insects as gifts to their predatory mates before approaching them. With the females distracted by food, the males don't have anything to worry about. They can go forth and copulate. Sometimes, that ploy fails. When their slow and cautious approach triggers their rapacious females' instincts, they draw in their legs and pretend to be lifeless.[3]

The female common hawkers faking death amazed and perplexed Khelifa. He had seen nothing like it in the ten years he had studied dragonflies. He recounted his observation to colleagues, and everyone laughed. 'They were both surprised and amused since the story is quite relatable,' he says.

He investigated whether this was a widespread behaviour in the species. When males chased thirty-five female common hawkers after they had laid their eggs, thirty-one fell out of the skies. Four didn't take evasive action and were forced to mate again. Despite lying still on their backs after plummeting to the ground, the males saw through the pretence of six others. Perhaps, the females' yellow and brown colours contrasted against their landing spots making them visible.

About 60 per cent of the observed female hawkers evaded their suitors, leading Khelifa to think this is a common behaviour in the species. When they crash-landed, they didn't suffer concussions or lose consciousness. They were alert even as they viewed the world upside down, and the researcher could catch only four while twenty-seven escaped. To them, sex-obsessed males were no different from hungry predators.

Sexual harassment is common in these hawkers as males outnumber females at nesting sites, says Khelifa. The females don't have male protection when they lay eggs. They are alone and vulnerable to bullying. For males, it pays to be coercive because they get more sex and the chance to father the next round of the females' offspring.

Khelifa held 50 per cent of the males patrolling two ponds captive for a day. With fewer marauders to escape from, the females grew bold, laying their eggs in open vegetation. And they didn't have as many encounters with males when they soared into the sky. A combination of female vulnerability and male-biased sex ratio creates the conditions for the evolution of this behaviour, suggests the researcher.

Is the cost of mating so high that female hawkers have to resort to such extreme measures?

'The function of the male penis is not only limited to sperm transfer but also to remove sperm left in the female's storage organ from previous matings,' says Khelifa. The penis could damage the females' reproductive tract while scooping out another male's sperm. 'One copulation is enough to fertilize all eggs. It is not in the females' interest to carry out multiple copulations per day.'

The breeding behaviour of the related southern hawker (*Aeshna cyanea*) is similar. Droves of males hover over nesting sites, and females seek sheltered spots out of their sight to lay their eggs. When males try to interrupt their egg-laying and drag them away, the females cling to plants instead of playing dead.[4] They also clap their wings and shake their bodies to signal their unwillingness to mate again. Other species bend their abdomens to make swift U-turns or fly in loops to evade pursuers.

Insects aren't the only ones to feign death to avoid sex. Reluctant female Iberian ribbed newts (*Pleurodeles waltl*) lie motionless on their backs too.[5]

This behaviour is prone to evolve in species 'where males use an aggressive approach of "conquering a female" by constraining them, not giving the females a choice,' says Sunita Leentje Janssenswillen, at Vrije Universiteit, Belgium, who studied the amphibians. When males of some salamander species 'politely' invite females by waving pheromones towards them, females didn't use this tactic. 'Also, it will be more likely to evolve in species that use faking death as a defence mechanism.'

This doesn't seem to apply to common hawkers since they don't play possum when threatened by predators. But other species of dragonflies do.

There may be another explanation for the sham performance. Females possibly judge the quality of males by 'testing' their persistence, says Alex Aguilar, of the

Universidad Nacional Autónoma de México and editor of the book *Dragonflies and Damselflies*.[6]

Female insects use various means such as wing colours, songs, and odours to gauge whether the males make good partners. Or, they may put potential suitors through their paces. Female black scavenger flies shake when males mount them and only mate with those that persist.[7] 'This possibility, I agree, is difficult to test but is the rival hypothesis in all cases where sexual coercion is invoked,' says Aguilar.

Crashing into bushes probably hurts the females, a high price to pay to avoid sex or test males. 'Given the speed at which they crash-land, there is a possibility that repeated crashes leads to physical damages, particularly in the wings,' says Khelifa. 'However, it was remarkable that females landed in vegetated areas so the crash-landing impact might be buffered by plants.'

If they pretend to be dead to choose males, then they may trade shorter lifespan for more offspring that would inherit their fathers' persistence, says Aguilar. This could work out in favour of females.

If faking death offers benefits, why isn't it more widespread?

Scientists don't know the answer since they haven't studied the sexual behaviour of most other species. The difficulty in catching, marking and observing the insects in the wild is a stumbling block. From the little they know, female common hawkers may not be the only dragonflies to act lifeless.

'I have observed that females of black saddlebags (*Tramea lacerata*), green darners (*Anax junius*), and giant darners (*Anax walsinghami*) stop flying and fall down to water or vegetation when taken by males,' says Aguilar. 'Females remain motionless, at times upside down. Possibly, these other species do not show elaborate behaviour such as moorland hawkers, but they definitely have similar responses.'

Whether the females play dead as a test or to avoid sex, most of the males flunk by leaving them alone.

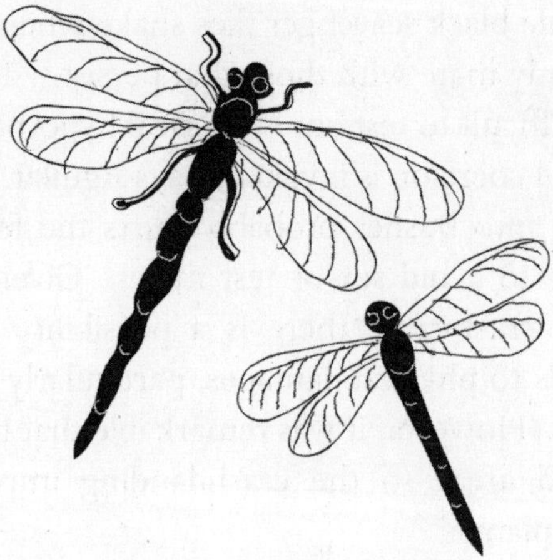

37

Scolding Egyptian Fruit Bats[1]

EGYPTIAN fruit bats (*Rousettus aegyptiacus*) spend the day roosting in large colonies inside caves. Instead of sleeping and conserving their energies for their nightly forays to feast on fruits in the surrounding countryside, they seem to be in a constant state of dissatisfaction, shrieking in unison with a few interludes of peace. To any human within earshot, their fractious cacophony sounds like nonsensical screaming. Scientists labelled these voluble animals' cries as aggression.

'Is this merely "shouting", or is there information amid the noise?' asked Yossi Yovel, of Tel-Aviv University, Israel. He and his students filmed twenty-two captive bats with infrared sensitive cameras and recorded their vocalizations using ultrasound microphones over the course of seventy-five days.

Each six-inch-long mammal with a two-foot wingspan wore a distinctive collar with a reflective disc for identification.

The scientists narrowed down 15,000 calls of seven adult females since the males didn't call much. After synchronizing the audio with video, they saw who called to whom, under what circumstances, and how the others responded. The bats squabbled all the time. Obviously, an argument was over a snack when one had a piece of fruit that the other tried to snatch. No prizes for guessing what a female was screaming about when a male attempted to mount her.

Four issues – food, sleep, mating, and space – bothered the animals. They didn't berate the colony at large but directed their scoldings at specific bats that upset them.

The interactions are almost always between two individuals that were often in physical contact with the vocalizing bat facing its peer, says researcher Yosef Prat.

Could the researchers guess the bone of contention by listening to the calls alone?

They used a speech recognition algorithm to pick out individual bat's calls. Machine learning is a radical tool in the study of bioacoustics, helping scientists pick out subtle differences among similar call patterns. The researchers 'taught' the computerized system using the audio and video recordings of bat squabbles. It then identified and classified patterns in 60 per cent of the animal cries from audio recordings alone. It even detected tone inflections as the bats addressed every individual differently depending on whether they were 'talking' to males or females, friend or foe. An eavesdropping bat can tell who is calling, with whom the caller is arguing, and about what.

'Our finding might be akin to a human speaker who uses varying intonation towards different listeners (e.g. male vs. female addressees) while using the same words,' the researchers write.

A batty harangue ended with one or both bats leaving, or both opting to stay. To their surprise, the scientists discovered they could predict how a squabble might play out even in darkness.

Does this mean that the bats have a vocabulary?

'I cannot rule out the existence of some vocabulary,' says Prat. 'We were not able to identify them in this study. But I believe that variations in the acoustics between contexts may be the evolutionary origin of vocabulary.'

The researchers feel if an algorithm can pick out subtleties in bat cries, the animals may be even more proficient at it. After all, the social animals live in dark caves where they can't see each other. So, their cries pack a lot of information.

When they began the study, the researchers suspected the bats conveyed some messages. 'But we had no idea how many details we could actually recover from this noise,' says Prat. 'In fact, we believe that we only discovered the tip of the iceberg, since we only used very specific aspects of the calls.'

In a later study,[2] the researchers studied whether these bats have dialects and how they learn them. They isolated fourteen wild-caught pregnant ones in three chambers, where each gave birth to a single pup. They piped the cacophony of a colony with 100 to 200 bats into the chambers. While those in one cubicle listened to normal bat calls, the rest heard artificially higher and lower pitched cries.

Although the pups paid attention to their mothers' voices for three months, they also took notice of the colony chatter. Their mothers' voices form a tiny percentage of what they listen in the formative weeks. In the wild, the young learn from the constant din of their cave mates with whom they can't interact or even see in the dark. When the captive pups weaned, the researchers released the adults and continued to channel the recordings into the enclosures.

As one-year-olds, the pups called in the frequencies of the recordings piped into their enclosures. Even the two groups exposed to unnaturally high or low tones matched the respective timbres despite their mothers' dialect being different. They are born with squeaky voices, and as they become adults, they lower their pitch under natural circumstances. The ones mimicking the high-frequency sounds would have had it especially hard as they went against their natural tendency to modulate their voices to the lower side.

Songbird nestlings pick up their repertoire from their parents. The social gathering and vocal nature of Egyptian fruit bats predisposes their young to learn from the entire colony, the first instance of crowd-learning in any animal. The very same conditions also create the need to tell their neighbours off.

This is perhaps the only language made up entirely of scoldings.

38
Music in the Wild[1]

MALE palm cockatoos (*Probosciger aterrimus*) fashion tools for only one purpose: to create a rhythm. Other versatile tool-wielders such as chimpanzees and New Caledonian crows employ implements to fill their bellies. This exceptional talent makes the palm cockatoos the only creatures besides humans to combine tool use and musical prowess.

G.A. Wood, about whom little is known, first reported the palm cockatoo's extraordinary behaviour in 1984.[2] Following the sounds of loud tapping, he observed one of these grey-black parrots holding a length of green stem in one foot. Wielding it like a mallet, the bird struck a dead eucalyptus tree trunk. On another occasion, he watched one using its large curved bill to break off a twig, strip off the leaves, and trim the ends before striking a tempo. The birds also gripped the round nuts of *Grevillea glauca*, called the Bushman's clothes peg tree, to pound hollow trunks.[3]

Their drumming ranged from two beats to more than 100. The longer they played, the more polished the stick became. Although ornithologists report the birds rapped on wood with a clenched foot or even their enormous beaks, the palm cockatoos' instrument of choice was a stem or nut.

Why do the parrots go to such lengths to beat a tattoo? Are they checking the durability of tree hollows before nesting? Perhaps, they attract the attention of females to potential egg-laying sites. Or, is the drumming a way of proclaiming their territories to rivals? Maybe they impress their mates. These questions remained unanswered for over thirty years since Wood's discovery.

Researchers from three universities in Australia studied the percussive skills of palm cockatoos at Kutini-Payamu or Iron Range National Park. Christina Zdenek, of the University of Queensland, spent the months from June to December of the years between 2009 and 2015 in this remote area.

Everything about the study was challenging, she says, including the primitive living conditions in a tropical humid forest. One year, she camped in an abandoned abattoir, and in a two-walled shed during another season. She had volunteer help for some time, but often she gathered data alone. She recorded 131 drumming sessions of eighteen palm cockatoos, no small achievement considering the birds' elusive nature.

Another difficulty was finding the birds without scaring them. Performing males sway and pirouette with outstretched wings while unfurling the slender crest feathers slicked back over their heads. Sometimes, they mix their soprano vocals with their drumming. Throughout this musical performance, their naked cheeks flush crimson in vivid contrast to their black plumage. Early in the study, when the birds had flown off to forage, Zdenek climbed up a wobbly single rope to measure the hollows. She also hoisted a ten-metre extension pole every few days to look into the tree hollow and gauge how far along the nesting process was. No matter how tired she felt at the end of the day, she organized her data, labelling and backing it up at least once. Given these challenges, it is not surprising that this is the first and only detailed study of this remarkable behaviour.

Back in the lab, analysis of the recordings showed the cockatoos maintain a regular and predictable beat. Each bird prefers a specific tempo, anywhere from fast allegro to slow lento. Some have more sophisticated tastes, pulling off an accelerando (speeding up) or rallentando (slowing down). The birds consistently perform to their preferred pace, and their neighbours may even recognize them from their distinct rhythms.

'These drumming displays, especially if performed with a hardwood stick on a hardwood, i.e., not live tree bark, can be heard up to 150 metres away,' says Zdenek. Over 70 per cent of the drummers played for an audience. The size of the listening group wasn't large. In fact, in every case, only one female cockatoo was in attendance. If she liked what she heard, she brought twigs to the nest hollow.[4]

'Male palm cockatoos thus appear to be more like solo musical artists or the beat setters of musical ensembles (for example, drummers in western rock bands) who have their own internalized notion of a regular pulse, and then generate the motor pattern that creates the beat,' write the authors.

Palm cockatoos build a cushion of sticks within tree hollows in which to lay a single egg. While carrying nesting material, one may have accidentally knocked on the resonant chamber and attracted female attention. The researchers surmise this is how a culture of drumming was born. Once the male succeeds in his overtures, the drumstick is no longer needed and he puts it to another use. He splinters it with his bill and adds to the nest.

Although the palm cockatoo ranges into the neighbouring island of New Guinea and smaller offshore islands, only the ones in Cape York peninsula are sophisticated drummers. In Papua New Guinea, males don't make a drumstick but rap with a closed fist.

Not all bachelor cockatoos even in Australia have a musical bent of mind; some have none. The researchers suggest playing percussion may be a cultural trait.

Since the performers crave female audience, how do others learn to fashion an instrument and beat it? 'Young males sometimes hang around, but only once did I see a young male present at a drumming event,' says Zdenek.

In his 1871 treatise *The Descent of Man*, Darwin suggested that the human proclivity for music and rhythm must predate our evolution. If this were the case, at least some creatures ought to share our innate fondness.

But few creatures can hold a rhythm. Aniruddh Patel, of Tufts University, found only two captive animals that could keep time to a beat.[5] Snowball, an Eleonora cockatoo, became an internet sensation for bobbing its head and moving its feet to tempo, and Ronan, a California sea lion, swayed to music.

Wild chimpanzees in Taï National Park, Côte d'Ivoire, beat buttresses of trees with their hands and feet while hooting,[6] but they don't stick to a steady pace. Chimpanzees tested in a laboratory had little success tapping in time.[7] One kept up with a metronome at one tempo but not when the speed varied. Similarly, bonobo Kuni tapped close to her preferred pace only.[8] But a chimpanzee called Barney drummed on an overturned barrel with elan.[9] While a few such extraordinary primates stick to a tempo, researchers haven't found a whole population of musically inclined animals or birds like the palm cockatoos.

Many features of the parrots' drumming behaviour remain to be researched. How do young males learn to beat? Do their styles change over time? Do drummers father more offspring than non-musical ones?

Percussion by palm cockatoos has a regular pace with the repetition of some bars like human-produced rhythm. In human societies, the social setting, drumming and dance form a package deal. But the parrots rap only to court their mates.

Despite these differences, the authors say the similarities suggest an evolutionary link as Darwin predicted.

Appreciative female palm cockatoos blush, bob their heads, sway to the beat, and display their flamboyant mohawk crests. Not all that different from humans.

39

Land Snails' Secret Weapons[1]

THE hard shell of a snail is an armour against predators and a refuge from dry conditions. Despite its utility, producing it is difficult on land where calcium is in short supply. By getting rid of this burden altogether, slugs became free to move into soils without the mineral, areas their shelled relatives had avoided. Snails have yet another essential use for the tough spiral structure on their backs.

Many species of roundworms, also called nematodes, live on snails and slugs. A pesticide company markets one *Phasmarhabditis hermaphrodita* as an organic natural slug-killer. Farmers and gardeners mix the worms, less than two millimetres long, with water and spray. The roundworms track slugs through their mucus and faeces and infest them, forcing their hosts to go off food and die. They then dine on the carcasses. Although hermaphrodita is relentless in its pursuit of slugs and snails, it has no effect on some others such as the giant African snails and leopard slugs. Needless to say, it is also not attracted by human beings.

While on vacation in 2014, Robbie Rae, of the Liverpool John Moores University, UK, walked his dog on the sand dunes in the north of Scotland. He saw heaps of pretty white-lipped snail (*Cepaea hortensis*) shells and took a handful to work. When he examined them under a microscope, he found small, squiggly structures on their inside walls. They appeared to be minor imperfections like cracks or scratches. But the more he looked, the more he became convinced they were nematodes.

'The snail shell, I thought, was a static structure, hard as a rock,' says Rae. 'My brain could not understand this at all. It just seemed so weird, so I didn't do anything with it and tried not to think about it.'

Later that summer, Alex Williams, one of Rae's students, carried out an experiment infecting brown-lipped (*Cepaea nemoralis*)[2] and giant African snails (*Achatina fulica*)[3] with nematodes. The infected creatures continued to gain weight and munch their way through the greenery. On dissecting the gastropods, as snails and slugs are also called, Williams found something surprising. When Rae entered the lab, his student shouted, 'They're in the shells! They're in the shells!' 'I didn't know what on Earth he was on about,' recalls Rae. Williams had discovered hundreds of worms trapped and killed within the spiral structures.

'The nematodes appear as if perfectly preserved in amber and are completely covered by unknown cells,' wrote the researchers.

In the latest study, Rae investigated whether land snails' shells performed additional duty as nematode-killers. He infected captive brown-lipped snails with hermaphrodita worms and observed the results by dissecting the gastropods. Cells in the shells perhaps sense various proteins or collagens on the nematodes' skin. Once they identified the parasites, they stuck to them and multiplied, fully engulfing the roundworms that fused with the inner wall of the shells. Oysters protect themselves from irritating particles of sand or parasites using the same strategy, exploited by humans for culturing pearls.

Brown-lipped snails cannot differentiate between harmful and harmless nematodes, plastering *Steinernema feltiae* and *Heterorhabditis bacteriophora*, species that don't prey on gastropods, to their shells.

By analysing DNA, Rae identified four species of roundworms from brown and white-lipped snails and a fifth one from common garden snails (*Cepaea aspersum*). He found entombed nematodes in brown-lipped snail shells collected by the evolutionary biologist Arthur Cain in the 1950s. With four species of snails proven to inter their parasitic assailants, Rae sought the big picture. He examined 1,321 shells of forty-three genera from twenty families from around the world, gathered between the late-nineteenth and mid-twentieth centuries. Of these, thirty-four species from fifteen families had killed roundworms. Even ones over 500 years old were gravesites of nematodes.[4] If this behaviour is so widespread, it must have evolved in a common ancestor before the different lineages split about ninety to 130 million years ago.

This was the hardest part of doing the study. 'I spent a lot of time trying to scratch away at the surface of the shells to get at the few nematodes which were encased in the shells,' Rae says. 'A lot of the time when I managed to bore into them they would just crumble away. So, there was a lot of failed

DNA extractions and a lot of wasted reagents. It was very frustrating.'

Slugs don't have a visible shell, but many species have reduced versions on their backs, inside their bodies, or even as granulated particles. Although they are unsuccessful against some nematodes, they use all these types of shells to kill others.

Rae plans to find answers to questions such as: How do shell cells recognize roundworms and how have some worms evolved to avoid them? Another avenue of study is the evolution of extinct species extracted from ancient shells.

'Nematodes are not well preserved during fossilization, but perhaps their DNA is in these snail shells,' he says. 'After all the encased nematodes would be sheltered from extreme temperatures and water which break down DNA.'

Gastropods belong to a large family of soft-bodied molluscs. Some members of the ammonite group, extinct nautilus-like marine creatures that lived approximately 400 million years ago, trapped parasitic flukes in the same way as land snails kill nematodes.[5] Palaeontologists call the structures with entombed parasites 'Housian pits', which everyone else calls 'blister pearls'. Modern-day marine molluscs, for instance bivalves, isolate flukes before they can do damage.

Gastropods, however, seem to kill nematodes only. Of the 5,000 shells Rae examined, none had evidence of other parasites such as flukes and flies. Across the developing world, flukes aided by their snail hosts spread diseases, such as schistosomiasis, affecting millions of people. Why don't the gastropods save humans from a lot of misery by going after these flatworms as their marine relatives do?

Flukes don't kill snails; they only hitch a ride on them for a part of their life cycle. So, Rae thinks there might be no need to mount an immune response against them.

Carrying their refuges on their backs seems like a drawback, but some land snails are using them to win the battle against their nemesis.

40

Sex Drive in Fish[1]

MALES that have a lot of sex make more babies. Their sons inherit their sexual prowess and become successful at procreating too. This lineage of high libido would have a clear advantage over ones with lower sex drive, outnumbering and replacing the latter entirely. But that isn't the case with mosquitofish. Not all males spend most of their time in pursuit of females. Some seem to be less horny.

Eastern mosquitofish (*Gambusia holbrooki*), a native of eastern US, was introduced throughout the tropics to control mosquito larvae. It belongs to a group of freshwater fish called poeciliids that have sex to produce live young. Its penis-like gonopodium, a modified anal fin, deposits sperm by hooking on to the females' vent.

Male poeciliids are probably the most ardent sex maniacs in the vertebrate world.[2] The light-coloured, four-centimetre-long male eastern mosquitofish doesn't court and cajole his larger mate; he harasses any female, by approaching her from behind and touching her genital pore with his snout. Fish biologists call this behaviour 'nipping'. Or he aligns himself below her and thrusts his gonopodium into her pore.

Sex is costly for female mosquitofish. When pursued by males, they interrupt feeding and go hungry.[3] The gonopodium, armed with claws, can injure the interior walls of the genital pores of the female mosquitofish.[4] While males may copulate multiple times a day, females do without for several days.

Carolin Sommer-Trembo, of the J.W. Goethe University, Frankfurt, Germany, and her colleagues sought answers to the question: what keeps male mosquitofish with both low and high sex drive in the gene pool.

Male fish not only differ in their sexual activity but also on a range of other traits such as body size, shape, colour and personality. How to ensure the females react only to the degree of male sexual activity and not any of these other factors?

'We found out that the combination of a cold beer and sitting there quietly just watching the fish is the best way to get new ideas for our studies,' Sommer-Trembo says.

She resorted to computer animated males nipping virtual females with different levels of persistence: low, medium and high. Along one wall of the testing aquarium was the screen. Now, which of these three intensities did the females find attractive?

She chose sexually experienced females from a mixed group of male and female fish and raised another group of females from eggs. This second group had not met a male yet and was sexually naive.

When sexually experienced females were alone while viewing the animated libidinous male harass virtual females, they stayed away from the screen. But they moved closer when they saw ones with medium sex drives. They even preferred the least ardent males to the oversexed ones.

Horny males can jeopardize the lives of lone females. The latter would make a tempting snack for a passing predator since they are larger than the males and distracted by their suitors. This is probably why solitary, experienced females evaded high libido males.

As a shoal, experienced females' preference for medium lust didn't change. But they were willing to consort with the most ardent males.

Prey fish seek safety in numbers to escape from predators. In the wild, the female mosquitofish put distance between each other when feeding, but the sight of a male makes them cluster together to avoid harassment.[5] This evasive behaviour gives them the confidence to mate with an oversexed male since they won't be the sole recipient of his persistent attention.

Inexperienced females weren't wise to the ways of the world, showing a marginal preference for horny males.

Since medium sex drive is most popular with females, high libido ought to be at a disadvantage. However, two factors play to the oversexed males' advantage. The willingness of virgin and shoaling females to mate with them and their higher harassment of females mean they score sex and produce offspring.

The question of why would a population maintain different behaviour has dogged the life sciences from the time of Darwin. Often, scientists struggled for answers. In this case,

Sommer-Trembo and her colleagues' experiment has come up with a neat explanation.

Although males seem to control females' sex life by bullying, the latter exercise their own choice and maintain variation in male behaviour. Females choose mates by weighing the costs and benefits of the circumstances. Since naïve ones expressed no choice, they must learn through experience, say the researchers.

Another study confirms this.[6] By studying the body shape of over 100 species of poeciliids, biologists at Kansas University say males and females evolve in different ways. While natural selection and the environment shape females, they in turn select traits in males.

If female mosquitofish know the costs of various male sexual behaviour, Sommer-Trembo and her colleagues wonder if they learn from watching others' mating encounters. Or do they learn only from their own experience?

No doubt she's sitting in the lab surrounded by aquaria right now designing an experiment to answer this question. With a cold beer.

41
Voting in African Wild Dogs[1]

EVERY member of a pack of African wild dogs (*Lycaon pictus*) has to agree to do the same thing at the same time. If a few want to sleep while the rest wish to go on a hunt, the first group would be left vulnerable and the second one may not be successful. How do they communicate their intentions?

At first, biologists assumed the alpha pair that rules the pack was autocratic in making decisions, and the other members had no option but to follow their leaders' commands. Then researchers observed a greeting ritual they call 'rally'. After lazing in the tall grass of the savanna, the restless dogs run towards each other with their heads held low, mouths agape, and ears folded back. Domestic dogs excited by the prospect of going for a walk behave in a similar manner. When the eager African wild dogs touch noses, the approaching dog licks and pokes the corner of the other's mouth while making high-pitched whines, whimpers, and twitters. Biologists thought this repertoire was a preamble to hunting.[2]

Researchers from the US, UK and Australia studied the behaviour of five packs totalling forty-nine dogs at Moremi Game Reserve in the Okavango Delta, Botswana, for a year between June 2014 and May 2015. They located the packs by tracking individuals wearing radio collars and identified each dog by its unique pattern of brown and black blotches. By viewing video recordings, they noted which animal started a rally and how the others responded.

They observed the packs don't set off on a hunt after every greeting ceremony. Sometimes, after a rally, the whole pack settled back for another rest. If the rally itself wasn't enough to shake the dogs' inertia, what could?

Amongst the wild dogs' vocal chatter, the biologists identified 'audible, abrupt exhalation of air through the nose, or sneezes'. After one sneezed, the whole pack appeared to catch the contagion. More than the rallies, the research team realized, the number of 'achoos' predicted whether the pack set off to hunt. In African wild dog parlance, a sneeze apparently says, 'Let's go hunting.'

'We couldn't quite believe it when our analyses confirmed our suspicions,' says researcher Neil Jordan, of the University of New South Wales, Australia. 'The sneeze acts like a type of voting system.'

The researchers couldn't definitely prove that sneezing was the signal to move. The thick grass obstructed their view, and they were not sure if the canids voted by some other means. Perhaps, the dogs sneezed to shake off their lethargy and clear their nasal tracts so they could pick up scent trails during the hunt. Since the chances of the pack departing on a hunt were higher after a communal bout of sharp exhalations of air, it may well be a way of reaching a quorum.

Playing recordings of this sneeze and gauging the dogs' reactions would prove if this sound, and not any other, urged the rest of the pack to go hunt, says Marta Manser, of the University of Zurich, Switzerland, who wasn't involved in this study.

Other canids have a similar repertoire of sharp exhalations: foxes 'pant' as an invitation to play,[3] coyotes 'huff' in alarm[4] and dingoes 'snuff' when nervous.[5] Unlike expressions of fear and anxiety, the wild dog sneeze doesn't disturb others in its pack. They don't look askance or are startled at the sneezer. These appear to be the only animals to cast a vote with a sneeze.

Not every African wild dog vote is equal. It is rigged in favour of higher-ranking members, and reaching a quorum is a tricky proposition.

If the dominant pair didn't go 'achoo', 'more sneezes were needed – approximately ten – before the pack would move off,' says the main researcher, Reena Walker, of Brown University, US.

But when the alpha dogs give the signal, they need a minimum of only three sneezes from the others to set their pack trotting. It's not clear if one dog gets only one vote. Would a restless dog sneezing a few times count as much as several dogs sneezing at once? The researchers say they didn't investigate which criterion was critical to pull off the vote.

Why do dominant dogs find it easier to rouse the pack than others? Meerkats also use a similar decision-making process, but they are more democratic. For one thing, they aren't cooperative hunters like African wild dogs. They call 'shall we move' when they are grubbing for beetles and scorpions in tall grass out of sight of each other. By the time one makes the first call, they've hunted out the area and it's time to move. So, it doesn't matter who takes that call since there are fewer delectable grubs to pick anyway.[6]

African wild dogs, however, hunt as a pack to bring down prey larger than themselves. The cooperation of every member is crucial for the chase. Hence the vote to decide whether to stay put and laze, or go out and bring down a gazelle. 'Sneezing occurs throughout the active social parts of rallies as well as through quieter part of rallies right before the dogs move off,' says Walker. 'We observed a peak of sneezing in the last minute of rallies, right before departure.'

Sometimes after a sneezing session, they lie down and go back to sleep. Presumably the call to hunt didn't receive enough votes. No doubt several minutes later when their stomachs growl, they'd reach a quorum of sneezes.

Irrespective of who initiated the vote to hunt, once they are on the move, the dominant ones lead.

Although they have a voting system, however biased it may be towards the upper echelons of power, it's an impressive feat for the dogs to count to ten.

42
Locking Poisons Out

THE poison dart frogs' bright blue, orange, yellow and red colours didn't evolve to entice children but to warn predators of their lethality. About a hundred of more than 170 species are toxic, but they don't synthesize their own deadly alkaloids. They absorb the substances from their insect prey and stock them in glands embedded in their skin. When a predator ignores the warning colours and grabs the amphibians, the pressure releases poison to the surface of the skin.

The indigenous hunters of Central America exploit these substances in bringing down birds and monkeys in the high canopy. They exhale a short powerful burst of air through a long hollow wooden tube, the blowgun, releasing a dart aimed at their quarry. They coat the sharpened palm rib darts with the skin secretions of poison dart frogs, among the most potent toxins in the natural world. As soon as the missile finds its mark, their prey tumbles out of treetops from the quick work of deadly substances.

The most poisonous species is the five-centimetre-long golden poison frog (*Phyllobates terribilis*), found only in the humid forests of Colombia. When in distress, it sweats a concoction of alkaloids, the most virulent being batrachotoxin (Greek for 'frog toxin') that blocks transmission of nerve signals. This causes respiratory paralysis or heart failure, or both, leading to the death of any predator foolish enough to ignore the warning colours. Every golden poison frog has about one milligram of batrachotoxin, enough to kill over 20,000 mice or fifteen grown humans. It is more toxic than curare and a thousand times deadlier than cyanide.[1]

The toxins don't affect the frogs themselves. Learning how these colourful amphibians evolved to resist their toxins without sacrificing normal functions could help design a new generation of painkillers as well as antidotes for smokers and drug users who want to kick their habits, say researchers. Two studies examined how the amphibians developed resistance to their chemical weapons.

Since the poisons are isolated within glands, could the frogs poison themselves? When predators bite them, alkaloids cover their bodies and get absorbed through the skin, says Sho-Ya Wang, of the State University of New York. If their attackers injure them, then the toxins seep in through the wounds.

The mere act of eating poisonous insects could be dangerous for the frogs, says Rebecca Tarvin, of the University of Texas at Austin. 'The chemicals (alkaloids) are lipophilic, meaning that they can pass through cell membranes and permeate throughout the body.'

Some species like the golden poison frogs live in groups of six or seven. During an occasional squabble, they can get a mouthful of unpleasant chemicals. Dealing with powerful toxins is an everyday reality for them.

Tarvin and her colleagues identified five unique amino acids in the muscles of the golden poison frogs.[2] Three evolved several times in the poison dart frogs' evolutionary history, but two mutated just once and only in golden poison frogs.

Sho-Ya Wang and her colleague introduced these proteins separately and in combination into rat muscle to test the immunity against batrachotoxin.[3] One of them – N1584T – blocks the poison's action and makes the frog species resistant to it.

Voltage-gated sodium channels open and close to let sodium in required amounts pass through cell membranes. Batrachotoxin attacks the amino acid that controls the gates, forcing the channels to stay open. Sodium ions flood the cells, causing nerves and muscles to fire nonstop. Substituting the amino acid with another changes the locking mechanism of the channels so batrachotoxin can't open them.

In another study, Tarvin and her team investigated the resistance of another species, Anthony's poison arrow frog (*Epipedobates anthonyi*) of Ecuador and Peru. This red or brown creature with yellow stripes produces a lethal alkaloid called epibatidine.[4] In low doses, the toxin has an opioid-like pain-numbing effect 200 times more powerful than morphine, and at high doses, it causes hypertension, respiratory paralysis, coma, and death.[5]

Instead of attacking sodium channels, epibatidine clings to nicotinic acetylcholine receptors, sabotaging the messages they send between nerves and muscles. For the frogs to develop resistance to their own lethal substance, the locks on the receptors ought to be changed to prevent the toxin from opening them just as N1584T does in the golden poison frog. But there is a problem with that solution. If the receptors are closed, acetylcholine won't have access to them and the normal functioning of the creatures is affected. How do the amphibians overcome this conundrum?

Tarvin and her colleagues sequenced genes of Anthony's poison arrow frog and its relatives, including some that don't store epibatidine. By comparing and ruling out similar amino acids, they zeroed in on four mutations in the receptors of not only Anthony's poison frogs but three non-epibatidine producing ones.

Instead of using rat tissue, the researchers introduced these proteins into human receptors to see the reactions to epibatidine and acetylcholine. One substitute, cysteine, in Anthony's poison frog locked out epibatidine as well as acetylcholine, like locking the house and losing the key. But other mutations compensated for this, giving acetylcholine access to the receptors.

It is possible that the golden poison frog's innate resistance to its alkaloid doesn't hinge on one mutation. Researchers from Colombia led by Roberto Márquez found N1584T in only one out of fourteen individuals.[6] They also discovered five additional mutations that might confer protection from the batrachotoxin.

The scientists don't have an answer for why frogs that have no use for epibatidine have the mechanism for blocking it. But they do know this invulnerability evolved thrice across the poison frog family.

'Was epibatidine more widespread millions of years ago?' wonders Tarvin. 'Is it costly to maintain resistance? Is it hard to "un-evolve" resistance once it is present? I also wonder whether other acetylcholine receptors in the frogs are resistant.'

Poison frogs aren't unique in deploying chemicals and neutralizing their effects on their own bodies. Toads of the Bufo family store bufalin in glands under their skin for self-defence. Bufalin messes with the functioning of the human heart. Two switches in amino acids work in combination to make the poison harmless to toads, says Kartik Sunagar, of the Indian Institute of Science, Bangalore. 'Interestingly, predators of bufonid toads have also acquired identical mutations in these receptors in a remarkably predictable and parallel fashion,' he says.

The fire-bellied snake (*Leimadophis epinephelus*) preys on the golden poison frog. How does it deal with the toxins? Scientists haven't studied the predator's immunity yet.

The indigenous hunters and their families don't have resistance to these poisons. But they have nothing to fear from eating the meat as cooking neutralizes the toxic effects. However, they have to take extreme precaution when handling their poison-tipped darts or arrows. If they accidentally inject themselves, there is no antidote.

43

Laziness Has Its Uses[1]

WORK never ends for industrious ants. From dawn to dusk, workers forage for food, tend the young, clean the nest and guard against invaders. But some do nothing at all. Daniel Charbonneau and Anna Dornhaus, of the University of Arizona, US, discovered that these freeloaders aren't one or two individuals or even in the minority.[2] At least 50 per cent of a colony laze, tripping the hardworking ones as they run about their tasks.

These ants, stouter than the busybodies,[3] stay put near the centre of the colony. 'They really just sit there,' says Charbonneau. If they move at all, they saunter,[4] doing a little brood care or grooming another worker.

When researchers removed active ants, the colonies replaced them.[5] But the productivity of the nest suffered. The substitutes weren't good at their jobs. The scientists speculated the replacements come from the ranks of these shirkers, who may not have the specialized skills needed to do any job well.

Charbonneau suggests the young are inactive as they are the most vulnerable members of the colony. Whether these jobless ants lay eggs is unknown. Their ovaries are operational because a chemical released by the queen hasn't suppressed their fecundity yet. In some species, they produce infertile eggs that become food for the rest of the colony. Although researchers suspected inactive ants had other uses, the proof was elusive.

Charbonneau and his colleagues studied why other ants put up with these goldbricks in their midst. They collected twenty colonies of *Temnothorax rugatulus*, a species from a high elevation pine forest in Saguaro National Park, Arizona.

'The main advantage of these ants is that they're very easy to work with,' says Charbonneau. 'You can collect whole colonies – workers, queens and brood – simply by flipping rocks and looking in boulder cracks. They're very easy to maintain in the lab in artificial nests and they're extremely hardy. As a result, they're fairly commonly used ant genera in research, especially for laboratory work. This is useful because we have some knowledge of their biology, and our results will be relevant for others working on these ants.'

Once the insects had settled in the lab, the researchers daubed paint in unique combinations on the head, thorax and abdomen so they could identify each individual in video recordings.

'Individually marking all of the workers (twenty colonies with ~100 workers per colony) and analysing the video data that we collected (tracking each individual ant of each colony for every second of multiple videos) was a time-consuming effort,' says Charbonneau. 'It was only possible because we had many undergraduate students who were participating in the project.'

The research team determined what jobs these rugatulus ants performed and distinguished between tired workers resting after their shifts and slothful ones.

They removed 20 per cent of the active workers from five colonies. In response, the remaining ants didn't increase their workload. Instead, in less than a week, others stepped in to pick up the slack. These new recruits came from the ranks of the idle.

The researchers removed 20 per cent of the lazy workers from nine colonies to see how that affected activity levels. Idleness decreased as the slothful ones grew busier. 'The top 20 per cent most inactive were significantly less "inactive",' says Charbonneau. But even after two weeks, none of the busy workers filled in for the missing shirkers.

The researchers then removed 20 per cent of randomly selected rugatulus workers, active and inactive, from six colonies. This disturbed the nests for a week. The most active ones didn't work as hard as they did before losing some of their mates, but the idlest ones didn't lift a leg to make up for the shortfall in the labour force.

The lazy ones weren't performing a necessary service, or they would have been replaced. The freeloaders were a rugatulus colony's B team, called into action when the experienced went out of commission.

This doesn't explain the presence of inactive ants in the ranks of all ant species, says Eisuke Hasegawa, of Hokkaido University, Japan. They could play other roles.

In one study of a Japanese species (*Myrmica kotokui*), the investigators had to remove all the active workers before the dawdlers swung into action.[6] Several busybody ants relaxed when researchers removed the lazy ones, as if the colony depended on maintaining a core group of idlers.

Hasegawa and his team came up with another explanation. They say workers react when the urgency level of getting a job done is high.[7] If others are already toiling at specialized tasks, the need to help them is low. For instance, egg cleaning and larval care are labour-intensive jobs. Hard-working ants performing them become tired and take a break. If no one tends the eggs, they could be infected with bacteria. As the danger of infection increases and reaches a critical threshold, it forces lazy ants to step up. While the dawdlers lower the efficiency of the colony, they are crucial to its survival[8], by enabling it to hedge its bets against inhospitable environmental changes.

Charbonneau agrees. Since a reserve labour force is 'not a widespread function, it suggests that there may be other underlying explanations for inactivity, and we don't yet know what those are.'

Lucky are the class of ants that specialize in doing nothing.

44
Shrinking Shrews[1]

WILD creatures have unique strategies for surviving temperate winters. Many birds cope by flying off to warmer climes. Some large mammals stuff themselves with food from spring to autumn, putting on enough fat to insulate themselves through hibernation. Shrews, however, can neither migrate nor hibernate. Keeping the internal thermostat on high drains energy at a time when prey isn't easy to find.

Typically, mammals in cooler, higher latitudes are larger than their warm-weather counterparts, a principle called the Bergmann's rule. Shrews turn this standard on its head, becoming smaller in the coldest parts of Europe and Asia. Two species even brave life in the Arctic Circle.

A more compact frame can survive on less. But the high ratio of surface area to body mass makes small bodies feel the cold much more keenly. The animals grow thicker fur to cut the loss of body heat, but they continue to burn fat reserves to keep warm. So, these small insectivores have an amazing solution – seasonally reduce body mass even more to lower food needs and save energy.[2]

Shrews don't downsize their physique alone. In 1949, Polish zoologist August Dehnel discovered their skulls in winter were smaller than in summer. Since then this adaptation takes his name, the Dehnel phenomenon. Non-hibernating mammals, weighing less than a kilogram, adhere to this principle.[3]

The two to three-inch-long common shrew (*Sorex araneus*), among the smallest Eurasian mammals, shrinks to a dramatic degree. It doesn't live to see more than one winter. Born in spring, it attains 70 per cent of its adult body mass by summer, gorging on earthworms. But it reduces in size over the following winter, when its favourite food burrows deep underground. The loss of weight can range from 9 per cent to 28 per cent, depending on which region of Europe it lives and the prevailing cold temperatures.[4] Their spinal columns shorten and the vertebral discs flatten. Most internal organs shrink, and metabolism drops. Even its brain, the most energy-expensive part of the body, shrivels between 10 per cent and 30 per cent in weight, reaching its lowest limit in March.

The following summer, when the soil crawls with earthworms again, the shrew regains nearly all that it lost.[5] Then it reproduces and dies before the second winter.

Once bones start growing, downsizing is unheard of in other creatures. When humans drop in weight, for instance, they lose tissue mass and bone density. But bones don't shorten nor do skulls shrink. As they age, the elderly become shorter as their vertebrae compress and arch, and they suffer a higher rate of fractures. Compacted bones are impossible to regrow.

However, the bones of astronauts who spend many days in space diminish in length and density. A year after their return to earth, they regain their original structure but don't recover their strength in that time.[6] So they exercise to prevent this loss from occurring at all.[7]

Shrews appear to suffer no consequences of shortening their bones. But there's a problem with the necropsy studies conducted so far. 'Each individual could be examined only once,' says Jan Taylor, of the University of Białystok, Poland. Such research projects face a serious drawback. It is possible, hefty individuals died in autumn and researchers caught the runts surviving into winter and jumped to conclusions.

Scientists would have to track the same animals in different seasons to demonstrate the waxing and waning of bodies. A group of researchers led by Javier Lázaro weighed and measured twelve wild shrews every month between June 2014 and October 2015.

Lázaro, of the Max Planck Institute for Ornithology, Germany, caught the worm-predators using wooden traps. The tiny animals cannot survive longer than two hours without food, so he checked the traps frequently. The researchers knew by appearance whether the tiny creatures were juveniles or adults.

'Animals in their first summer are easily distinguished from second-year animals since they have shiny fur and straight tails,' says Moritz Hertel, another researcher in the study. 'These animals are very territorial, males as well as females fight a lot. Older animals in their second summers are easy to recognize by their scars and general body condition.'

Lázaro anaesthetized the shrews to take precise X-ray measurements of their skeleton and braincase. Before release, he fitted them with microchips for identification. Their high metabolism made even the brief period of anaesthesia challenging, says Hertel.

Lázaro had to capture them again the following month 'in competition with all the owls, foxes and cats out there, to get the next data point,' says Hertel.

All the catching and measuring paid off. The researchers found the average body mass dropped 17 per cent in autumn and shot up by 83 per cent in spring. The skull lost an average of 15 per cent that regrew by 9 per cent. This is the most extreme case of reversible change in total brain size in a vertebrate.

'That something as rigid as the bone of mammals is able to undergo such profound changes is amazing,' says Hertel. 'We know little about the costs of shrinking the bones, brains and other organs and later regrowing them.'

Why don't young shrews grow to their winter size and then reach full adult stature as the weather warms up, following an incremental growth plan like average mammals?

'Bear in mind that evolution does not follow logic,' says Hertel. A life form undergoes random changes to survive in its environment. There is some speculation that bones could function as a kind of calcium reservoir to tide over a calcium-poor winter diet.

Juveniles may need large brains to set up territories and find shelter before the cold weather arrives.[8] Two of the many brain regions that shrank were the hypothalamus and neocortex, crucial for memory, learning and understanding the lay of the land.[9] Since the shrews don't move much in winter, neither confronting rivals nor courting mates, the downsizing probably didn't hurt. By summer, the animals expand their domain in search of food and mates, and the hypothalamus grew again, perhaps to accommodate the seasonal needs. But the neocortex didn't regain its size, and the researchers don't know why.

Doesn't the smaller brain size cause cognitive decline?

To test the mental abilities of shrews in summer and winter, Lázaro and his colleagues caught them from the wild and brought them to their laboratory.[10] They watched how far the animals walked to find mealworms. In the first and second summers of their lives, shrews found short paths to the food. But in winter, they wandered farther over the experiment enclosure before reaching their targets. The researchers associated this poor performance to the small size of the hippocampus. By shrinking their brains, the shrews somehow benefit from the trade-off between saving energy and deterioration in brain functions.

The researchers are still on the case of this enigmatic mystery.

45

Wading Birds in the Desert[1]

Banded stilts (*Cladorhynchus leucocephalus*) breed in such difficult conditions, it is a miracle they haven't become extinct. These waders flock in large groups at their regular haunts, the shallow coastal wetlands of southern Australia, catching insects and probing the soft mud for tiny crustaceans and molluscs. These 20-inch-high birds look for all the world like normal shorebirds – their breeding behaviour is anything but.

Once in a while, as if on a secret command, the waders forsake the coast and take to the skies, flying hundreds to thousands of kilometres inland. Where they went to breed was a mystery that rankled ornithologists for at least a century. With little evidence, they assumed the stilts, like many other birds, flew to Siberia.

In July 1930, a letter to a curator at Western Australian Museum, Perth, changed that premise. Mrs B.E. Cannon from Kukerin, Western Australia, wrote she had seen a rookery, thousands of stilts strong, on a sandspit in Lake Grace, about 250 kilometres southeast of Perth. She sent a blown egg and photographs as evidence. This was the last area bird aficionados expected to find the waders. Instead of flying to Siberia, the birds opted for another harsh place, the Australian desert.

In the following January, J. Neil McGilp and A.M. Morgan, two members of an ornithological association, saw a breeding colony of 54,000 stilts on an island at Lake Callabonna, about 2,000 kilometres due east of Lake Grace.[2] The nests, shallow scrapes in the ground, pockmarked the lakeshore without a bush to shade them from the scorching sun.

Rains sweeping over the desert fill the gypsum-crusted lake beds, beckoning the birds from afar. These downpours aren't seasonal or even predictable. Often, they fail, causing droughts. Nor are these lakes perennial. When the water level rises, dormant eggs of the brine shrimp hatch, creating a rich salty soup amid inhospitable terrain.[3] The short period of plenty fuels banded stilts to nest en masse. Since breeding is contingent on erratic and rare weather, the species' survival can be precarious.

Reece Pedler, at Deakin University, Australia, became fascinated by the birds' behaviour when he lived 'in a remote mining town in the Australian outback near some large salt lakes, where the banded stilts had bred previously.' He and his colleagues tagged fifty-seven of them with solar-powered satellite transmitters.

'When logging in to the tracking website each day to download our tracking data, we were at times astounded by the long-distance flights the birds had made overnight,' says Pedler.

Within a week of rains hitting distant salt lakes, eighteen birds flew inland from the coast, undertaking journeys between 150 and 2,300 kilometres. One travelled 2,200 kilometres in a little over two days.[4]

'The birds were moving much further and faster than we had expected,' he says.

Banded stilts wing their way straight to the area and don't stumble upon the lakes by accident. How do they know when it rains hundreds of kilometres away? Studies suggest they may hear the low-frequency sound of the showers, or even smell them.[5] They may sense temperature and pressure changes.[6] Large rainfall events may be easier to perceive, but it is astonishing they are aware of short spells of rain.

As soon as rains hit the interior, the researchers flew twenty-one sorties in a light plane over fourteen salt lakes in South and Western Australia. In six years, they found nine stilt nesting aggregations and recorded thirteen reproductive events. By using aerial photographs, they estimated between 200 to 40,000 nests at these sites. They approached seven colonies on foot to get data on nesting area size, density, the number of eggs in a nest, and egg measurements.

Each 250-gram female laid about three eggs, some produced as many as five, which were between 50 and 80 per cent of her weight. In comparison, the eggs of a close relative, the black-winged stilt, ranged from 32 to 66 per cent of the mothers' weight.[7]

'The logistical challenges in reaching some of the nesting sites on remote salt lakes were extreme,' says Pedler. After spotting them from the air, the researchers drove many hours in 4WD vehicles across rough tracks and walked through ankle-deep, or sometimes knee-deep, sludge to reach the nests.

Reports from Lake Torrens, 400 kilometres due north of Adelaide, South Australia, suggested the stilts nested only during major floods and ignored the minor ones. Ornithologists noted the birds raised young only thirty times in the following eight decades since Cannon first recorded the breeding event. Pedler and his colleagues built a twenty-nine-year record of lake inundation events using remote sensing data and found the water bodies flooded on at least fifty occasions. But the waders bred only on half those instances. If cloudbursts are rare in the outback, why do banded stilts forgo these opportunities?

The researchers discovered brimming lakes weren't enough to trigger nesting. Stilts preferred winter rains, even if minor, because the water bodies held water longer than heavy summer downpours. For instance, during the winter of June 2013, 9,000 pairs had between 12,000 and 14,000 young on Lake Torrens that had only 11 per cent spread of water. They didn't rely on rare major floods that strike once or twice a decade. Instead, by depending upon partly flooded lakes, banded stilts bred seven times more often than ornithologists had realized.

Judging if the lake's waters would last twelve weeks, until chicks fledge, was tricky. The birds frequently miscalculated. Despite the strenuous lengths the birds went to have chicks, they abandoned them before they hatched in seven of those breeding attempts. If the lake dried up and the young hadn't fledged, the flightless young trekked over the desert, crossed roads, and bumbled into townships in a futile search for other lakes.

A fast receding coastline is just one of the perils. Silver gulls (*Chroicocephalus novaehollandiae*) are another. The researchers speculated these predators of stilt eggs and chicks might be to blame in most cases of abandoned nests. Over the past eighty years, their numbers increased at many salt lakes in southern Australia, where the stilts are easy targets.

For instance, in 2012, gulls ate twenty-nine eggs an hour at Lake Eyre South, and when chicks hatched, silver gulls killed four every hour. Less than 5 per cent of the 30,000 eggs survived. Gull numbers were so high at Lake Torrens that the state environment department culled them.

'We don't yet fully understand why silver gulls have become a problem to the nesting banded stilts,' says Pedler. 'However, it seems that these gulls, which are native, have increased in number due to human waste management practices.'

These ground-nesting stilts probably fly farther to gull-free waters of northwestern Australia. But even these remote spots may not be safe for long. Fertilizer companies plan to mine these salt lakes for potash.

The stilts go to great lengths to have offspring, undertaking strenuous journeys deep into the desert and investing in hefty eggs. Between the silver gulls and erratic weather, those eggs have a measly 30 per cent chance of hatching. The odds of those chicks fledging are worse.

Within twenty-four to forty-eight hours of emerging, the offspring follow their fathers to the lakeshore where they feed. Since others hatch at the same time, the young band together to form creches, no longer dependent on their parents. Free from providing care, the adults try to squeeze in another round of eggs if the lake still has water. Once the frenzy of reproducing is done and they are back in the coastal wetlands, they can at least take it easy until it rains again in the desert. That won't be anytime soon.

46

Bird Nannies[1]

YOUNG male moustached warblers (*Acrocephalus melanopogon*) are the saviours of single mothers. If the females' consorts disappear, these bachelors step in, incubating eggs, feeding chicks, and chasing predators away.[2] Why do they forgo their own chance of breeding and help raise another male's offspring?

These warblers are not exceptional. Other species, like fairy-wrens (*Malurus spp.*), live in groups with relatives and unrelated individuals. When partners die, the survivors hook up with others, who nurture the nestlings.[3]

The olive-green coloured bell miners (*Manorina melanophrys*) assist their own kith and kin.[4] Instead of leaving their parents' territory after fledging, the young adults stay to help. Aiding one's own relatives, called kin selection, benefits parents who don't have to work as hard. By saving energy, they nest again.[5] The nestlings lack for nothing with the extra attention. Solicitude is a trait in 9 per cent of birds that nest in cooperative groups.

At one time, ornithologists assumed all helpers were related to the breeders. Molecular testing showed non-relatives regularly played surrogate parents[6] in about 20 per cent of these cooperative nesting species.[7]

In 2013, evolutionary biologist Christina Riehl, at Princeton, US, reviewed how and why these altruistic birds offer their services to breeding adults. While babysitting, young male pied kingfishers (*Ceryle rudis*) try to sneakily mate with the dominant female. If that ploy fails, they hang around as helpers, hoping to father the next round of offspring.[8] The female is more likely to accept a helper as a suitor if he assists with her earlier batch of nestlings.

But not all surrogate parents of unrelated chicks gain such clear payoffs. In fact, they could pay a price. The assistants forgo food while incubating eggs and foraging for chicks, losing weight and body condition. They aren't in the best shape to reproduce, and they could even jeopardize their own survival. What makes them act in such a generous manner?

According to one theory, in return for being allowed into the breeders' territory, parental aides 'pay rent' by helping with childcare. Another says raising others' nestlings allows them to practice skills they would need later in life. A third theory suggests birds are programmed to stuff food down the gaping mouths of any chicks, whether or not related. Other inferences include a shortage of females and the formation of alliances.[9]

Evolutionary biologist Sjouke Kingma, of the University of Groningen, The Netherlands, has studied cooperative breeding in birds since 2010 and found an overarching theory to explain surrogacy.

Birds need territories to meet their needs of food, water, shelter and nesting sites. He examined forty-four species that provided parental help, of which sixteen didn't have a problem establishing a territory. The non-breeders among them nurture others' chicks in the colony if they themselves fail to breed or they haven't matured yet.

For the rest, real estate assets were at a premium. When the habitat was already carved up by established pairs, their only chance of reproducing was by inheriting their fiefdoms in part or whole from breeders. Not all were rewarded. Only 30 per cent of the subordinates received a bequeath. This incentive drove parental aides to keep their wards well-fed, overlooking the distinction between relatives and strangers. They worked themselves to the bone, striving 50 per cent harder than caregivers that don't suffer a space crunch.

'Birds see their territory in the same way as we see our house,' says Kingma. 'My research reveals that the "home-owners" get much more help if the helpers stand to inherit their territory in the future. After all, you would be much more inclined to help someone maintain their home if you thought you would inherit it one day.'

How would birds know they stand to inherit the territory? How do they strike the deal?

When single birds with no prospects of breeding benefit from providing help, natural selection chooses that behaviour. It is an arrangement that favours all parties. The parents have extra help in raising young, and the nestlings get better care and thus survive to adulthood. The caregivers are paid in property later.

Although the idea that birds might trade childcare services for territory is not new, 'nobody really showed empirical evidence for it on this scale,' says Kingma.

The perks from providing support probably play an even greater role than is usually assumed, says Riehl. But this doesn't undermine the importance of helping relatives. 'Kin selection and direct benefits aren't mutually exclusive hypotheses. Rather, they often reinforce each other and both are important drivers of cooperation.'

In many species, property titles are dearer to males. It is possibly no coincidence that most surrogate parents are males. Territory is essential for the females of some species. Would they act as au pairs too?

Indeed, female white-winged trumpeters (*Psophia leucoptera*) assist a breeding female whose territory they will inherit when she dies. Since they are kicked out of their homes by their parents and cannot set up their own fiefdom, assisting a dominant female is the only way they can take a shot at reproducing.[10]

'Males and females would equally benefit,' says Kingma, 'although we see that helpers are more often male than female in many, but not all species. I think that the role of kin selection is slightly overestimated. Helping unrelated individuals may bring lots of benefits in the future.'

The complexity of this behaviour hasn't been fully unravelled. For instance, what determines which particular helper inherits territory? When many helpers support a pair, how do they establish their place in the hierarchy? How do they resolve property disputes?

Seen through the prism of real estate shortage, nannies among fairy-wrens and moustached warblers have struck a deal. When they inherit territories, they also get the bonus of a crop of aides to help them with their own parental duties.

47

Dads vs Cannibals[1]

STUDYING frogs in the field is like watching an art film unfold, says K.S. Seshadri. 'For aeons, nothing happens. Then you blink and miss one crucial scene that tells the whole story.'

The researcher and his colleagues from Ashoka Trust for Research in Ecology and the Environment (ATREE), Bengaluru, stood motionless one night as a white-spotted bush frog (*Raorchestes chalazodes*) squeezed into a small hole in a bamboo stalk. It wasn't hiding from the humans. Nor was it an exception. Seshadri observed other frogs of the species disappear into these portals. After many nights, he heard them call from within the hollow stems.

The researchers didn't know what to make of this behaviour. The species had come into the spotlight in 2011. Seshadri had been a novice when he joined experts to rediscover the two-centimetre-long frog for the first time in over 135 years. Drawn by an unfamiliar rapid staccato call at night, they found the caller. A luminous green frog with yellow starburst eyes sat on a bamboo culm in Upper Kodayar in Kalakkad Mundanthurai Tiger Reserve, Tamil Nadu.[2] The excitement of beginner's luck goaded him to research frogs, and what better species to study than the one he had helped to reclaim from obscurity. The tiny amphibian's bizarre behaviour foxed him.

A related species, the ochlandra reed frog (*Raorchestes ochlandrae*) in Calicut district, Kerala, sticks its eggs inside the hollow internodes of reed bamboo.[3] Frogs typically lay their eggs above or in standing water. The unusual breeding behaviour of the ochlandra reed frog didn't stop there. It belongs to a family of bush frogs that don't hatch into aquatic tadpoles that metamorphose to terrestrial adults. Instead, it emerges from eggs as fully formed froglets.

Since white-spotted bush frogs wriggled into an internode with determination and called from within, they probably reproduce like ochlandra reed frogs. There was only one way to find out.

Seshadri sliced open the bamboo culms and found eggs glued to the interior walls.[4] At the start of the reproductive season in June, males searched among stands of reed bamboo (*Ochlandra travancorica*) that grew along stream banks. They called, ballooning vocal sacs in their throats acting as amplifiers, when they found bamboo sections with holes. The males continued croaking after entering the tubular chamber. Attracted by these sounds, interested females crawled in with them.

Seshadri realized not any internode hollow with an opening would do. If the hole was at the top, then rainwater dripped inside and drowned the eggs. Best were the ones with holes at the base.

To get a better understanding of the species' breeding behaviour, Seshadri peered inside the hollow stems with an endoscope. He spied forty-three clutches of eggs stuck to the top of the column at thirty-two sites and monitored them daily. Since the eggs didn't have a shell to hold moisture in, they could dry up, killing the developing froglets. The 60-centimetre column of the internodes was cooler and more humid than the outside, providing ideal conditions for the embryos to develop.

Seshadri also realized the males were single parents. During the day, the fathers slept next to the eggs with their eyes half-closed. At night, when he poked the equipment into the bamboo, most of them remained where they slept or sat on the eggs. Some stopped calling and hid behind the eggs or at the base of the internode. A few brave ones tried to scare the endoscope by calling or attacking.

How critical were the fathers to egg survival? Predators like katydids and snakes were too big to enter the tiny slit in the internode. Seshadri caught the guardian males of thirteen sites and released them on the other side of the stream, leaving the eggs with no protection. He compared the fate of this group with another lot of thirteen sites which were protected by the fathers.

In less than three weeks, the fatherless egg masses faced rough times. Ants, flies, and fungal infections killed some. But the main threat to the eggs came as a surprise to Seshadri. Other male frogs wriggled into the hollows and ate 70 per cent of the ill-fated eggs. Only 34 of 129 orphaned eggs hatched.

Did cannibalistic frogs then serenade females and tend their own eggs?

'We did find the cannibalistic males in the usurped internode,' says Seshadri. But after he captured them to take measurements, they abandoned the stems. 'We did not really observe them attracting a mate and attending to eggs after cannibalizing an existing clutch.'

Compared to the dismal hatching rate of fatherless egg masses, froglets emerged out of all the eggs that had a guardian. Only one rolled down and drowned in rainwater collected at the base of the internode. Other frog parents snapped up flies and ants and ate infected eggs before they contaminated others. Perhaps white-spotted bush frog dads maintain the health of their eggs in the same manner.

Did this effort come at a cost to the males? They could be out feeding and mating during the time they spent guarding the eggs. But these bush frogs didn't seem to suffer much from lost opportunities.

They hunted every evening before returning to egg-sitting duties. Since they were particular about what kind of internode made a good nursery, ideal spots were in short supply. A male in possession of one was a catch, even if he was already tending a batch of eggs. He sang to advertise his asset, wooing females to spawn in his lair. He didn't have to abandon his guarding duties to go looking for a mate.

For every male that guarded eggs, two to three others hung around with no nest sites of their own. Fathers called incessantly through the night to ward off these rivals with cannibalistic tendencies. When such prized locations fell silent after the researcher removed the fathers, the other males probably realized the hollows were theirs to occupy. Besides gaining a breeding spot, they got nutrition by eating eggs.

The parental devotion paid off when an average of thirty-seven days later, froglets hatched. They hung around their natal spot for up to a month, absorbing the yolk in their bellies. Their fathers remained with them but didn't seem to take an interest anymore. Eventually, they also lost their possessiveness over their prized bamboo stem territories. Seshadri couldn't find any of them in December and January, outside the reproductive period. So far biologists know of fifteen locations in a 250-square-kilometre area where white-spotted bush frogs live.

It's not clear if the possession of a nest site makes devoted fathers and lack of it gluttonous cannibals. Did fathers protective of their own eggs gobble up others'?

In true art-film-enthusiast fashion, Seshadri will have to pay close attention to white-spotted bush frogs to answer these questions.

48

High Altitude Survival[1]

BAR-HEADED geese (*Anser indicus*) fly at elevations that defeat most other birds. They raise goslings in Mongolia and Tibet, and come autumn, soar over the Himalayas and down into the Indian subcontinent. They take several days to complete their migration, but cross the lofty mountain range at night in one marathon flight lasting seven hours. It may be easier to keep going once you are at that height rather than land and take-off again and again, but it poses many challenges. On their return journey to the north, they go from sea level to Himalayan heights, the steepest migratory flight of any bird.

Although birds make flying seem effortless, it sucks ten to twenty times more oxygen than when they are at rest. But this critical element is in short supply at high elevations, with only one-third to half the oxygen levels at sea level, causing a condition called hypoxia.

How do bar-headed geese sustain their passage at 5,000 or 6,000 metres? Scientists even recorded one flying at 7,290 metres.[2] The birds exploit updrafts of wind and glide once they reach cruising height.[3] But they cannot coast indefinitely when gravity pulls them down. They have to beat their wings to regain elevation and maintain it. In addition, the geese battle freezing nocturnal temperatures while soaring high above the mountains. Perhaps, the cool nighttime air contains more oxygen and regulates body warmth better than the heat of the afternoon. With all this exercise, they may get dehydrated.

Despite their plump build, bar-headed geese are extreme athletes, flying marathon distances without adequate oxygen or water in the cold heights.

'They endure significant hypoxia during the breeding season of three to four months [in Mongolia and Tibet], extreme hypoxia during migration lasting one to two weeks, and minimal hypoxia in the non-breeding season of six to nine months [in India],' says Sahas Barve, of Old Dominion University, US.

To overcome the mortal danger of flying through thin air, the geese have a suite of special features. Their lungs are larger than those of lowland birds of similar mass. A dense network of blood vessels in the wing muscles carries oxygen-rich blood to the extremities. They hyperventilate at seven times the normal resting rate without passing out.[4] All these qualities and more allow them to extract the most mileage from every breath of air.

Are birds that transit through high elevations built differently from resident ones? Unlike Asian bar-headed geese, Andean geese (*Chloephaga melanoptera*) and crested ducks

(*Lophonetta specularioides*) spend their lives in the mountains of South America. For one thing, the lungs of the Andean geese are larger than those of their Asian cousins.[5] Ten researchers from five countries led by Sabine Lague, of the University of British Columbia, Canada, compared how these species obtained more oxygen when it was in low supply.

Instead of working with captive birds at low elevation laboratories, the team chose to work with wild ones at high altitudes. In China, they caught and tested seven bar-headed geese, and in Peru, seven Andean geese and six crested ducks. They fit the birds into a flexible cradle and reduced the amount of oxygen by increasing nitrogen, simulating conditions at 4,500, 7,000, 8,500, and 11,000 metres altitude. They measured the birds' oxygen use, heart rate and blood pressure among other parameters.

The birds maintained an almost constant consumption of oxygen, even as the air they breathed had less of it. Bar-headed geese panted to get more oxygen into their lungs. But hyperventilating for long stretches causes carbon dioxide levels in the blood to fall, turning their blood alkaline. This doesn't affect birds as much as mammals. In fact, it seems to improve the lungs' absorption of oxygen.[6]

'They can extract up to 90 per cent of the oxygen from air and this ability goes up as the air becomes more hypoxic,' says Barve, who studied the effects of low oxygen levels on Himalayan birds but wasn't involved in this study.

However, the South American birds didn't breathe hard or take deep breaths. Their extra-large lungs were extremely efficient at extracting oxygen.

Bar-headed geese pumped twice the volume of blood by increasing their heart rate. But the Andean birds achieved the same effect by pushing more blood, high in haemoglobin, with every heartbeat. Either way increases the quantity of blood, which transports oxygen from the lungs to body tissues.

The authors suggest that bar-headed geese's adaptations of panting and rapid heartbeat work well for short durations at high altitudes, but they require a tremendous expenditure of energy. Hyperventilating and racing hearts aren't sustainable every day. Instead, year-round mountain residents Andean geese and crested ducks spend far less effort because of their efficient lungs and hearts.

The researchers compared the performance of the birds with people living in the Andes and the Tibetan plateau and found many similarities. Tibetans have normal haemoglobin, and they ventilate rapidly, like bar-headed geese. Every breath inflates their larger-than-average lungs with more air.[7] Their blood has double the levels of nitric oxide that dilates red blood cells, perhaps enhancing circulation and moving oxygen.[8]

Instead of inhaling more air to gain more oxygen, Andeans have higher levels of oxygen-rich haemoglobin like the Andean geese and crested ducks. Their nitric oxide levels are normal but the density of red blood cells, called haematocrit, is high. Increasing the number of blood vessels turns blood viscous, slowing its flow. In extreme cases, the thick fluid cannot carry oxygen to tissues, causing chronic mountain sickness. As much as 15 per cent of the people living in the Andes suffer from the disease, which causes pregnant women to endure high blood pressure and newborns to weigh less than normal.[9]

Despite living at high elevations for thousands of years, some adaptations work against the Andean people. But the birds sharing the mountains with them avoid this problem.

49

Firehawks[1]

WILDFIRES attract predatory birds. Every year 30 to 50 per cent of northern Australia burns. Dry savannas, with their high load of inflammable material, are tinderboxes. Scores of insects, reptiles, and rodents escaping from the conflagration are easy pickings. Once the birds snatch a morsel, they perch somewhere safe and tuck into it. Sometimes, they pick up barbecued prey, even when it's well-done. Being airborne, they aren't worried about getting singed by the fire, and they move with the fiery front.

Can these raptors make the leap from exploiting an already raging blaze to creating one? Evolutionary biologist Richard Wrangham proposes in his book *Catching Fire: How Cooking Made Us Human* that the one thing distinguishing humans from animals is the use of fire. Big brains require turbo power, and raw meat doesn't provide enough energy. But tubers are plentiful. Cooking them not only makes them palatable but also released brain development of ancient humans from calorific restraint. Manipulation of fire is a skill no other animal has shown. Not until now.

The Aboriginal peoples of Australia ignite the landscape at the end of long dry seasons, when wild food is hard to find. Burnt plants and grasses put out new growth that attracts herbivores such as kangaroos. When animals gather in numbers, hunting them becomes easier. Indigenous communities use fire so extensively that ecologists call their style of land management 'fire-stick farming'.

In his autobiographical book *I, the Aboriginal* published in 1962, Phillip Waipuldanya Roberts describes seeing a bird pick up a flaming stick from a fire front, and carry it with its talons only to drop it on unburnt dry grass half a mile away. Roberts suggested his people probably learned this technique of using flames to flush out prey from the birds.

Such stories are familiar to the Aboriginal communities living in Western Australia, Northern Territory, and Queensland, where people reenact them as part of their religious ceremonies. The emblem of the Mimal Land Management organization in the Northern Territory is an illustration of one such bird.

Reports of this behaviour intrigued Robert Gosford, an Australian ornithologist and independent researcher who is interested in traditional knowledge. He and his research team collected published records and put out the word asking for eyewitness accounts. They narrowed down a dozen ranchers, firefighters, and two academics, who claimed to have seen birds igniting fires. They shortlisted twenty instances from a 2,400 kilometre stretch of northern Australia, where raptors carried a smouldering stick and dropped it elsewhere. The birds, whom Aboriginal peoples call 'firehawks', were of three species: the black kite (*Milvus migrans*), the whistling kite (*Haliastur sphenurus*), and the brown falcon (*Falco berigora*).

Anthropologist Kim Akerman described how a black kite created a new fire about fifteen to twenty feet away from a blaze in Kimberley, Western Australia. This 1977 event wasn't a one-off. He saw the same action repeat itself in 1980. Anthropological linguist Denise Angelo not only heard these tales from the Aboriginal peoples but also witnessed the behaviour several times in the Northern Territory.

Dick Eussen, a firefighter and one of the researchers, saw a kite start a fire across a Queensland road by dropping a smoking stick on flammable grass. He says he put out seven fires ignited by raptors in Northern Territory. The birds make controlling wildfires even more dangerous.

One firefighter said that a few raptors set a small valley ablaze. Another reported they kindled a fire that got out of hand and partly burnt down a cattle station. Workers at another ranch killed firehawks with shotguns to protect livestock and property from going up in flames.

Many non-Aboriginal people initially dismissed these stories when they first heard them but later saw the behaviour for themselves. Some cursed the birds' ability to rekindle fires that had been stamped out.

'It's a wonderful story,' says Suhel Quader, of the Nature Conservation Foundation, Mysuru. 'The eyewitness accounts are compelling, but there are surprisingly few [of them], given the frequency of fires in the region plus the number of people out there trying to manage the fires.'

Perhaps, in sparsely populated northern Australia, observers were few. Firefighters work long hours putting out a blaze and can't watch birds at the same time. And the behaviour isn't widespread. Eussen reports that of twenty-five kites hawking for prey, only two attempted to pick up blazing sticks. One inept bird repeatedly dropped a smoking twig before it could carry it farther.

Another firefighter and researcher, Nathan Fergusson, observed such events at least a dozen times in his career. He says that out of hundreds of kites only a few picked up flaming twigs, and he now anticipates this avian behaviour in managing fires.

However, raptors flock to fires throughout the world. Large numbers of black kites wheel around the wildfires of Asia and Africa, yet there are no tales so far of this fascinating behaviour from other countries. Other birds such as buzzards also hunt at fire fronts, but there are no reports that they start fires. Maybe people haven't noticed, or nobody asked resident communities about them.

Perhaps, the northern Australian population of raptors alone has the cultural tradition of using fire-sticks. This is not unusual since chimpanzees use specific techniques of fishing ants (see Chapter 7: Conservative Chimpanzees). Of the different populations, palm cockatoos of Cape York alone drum with a tool (see Chapter 38: Music in the Wild).

'At present, we have scattered anecdotal reports of similar behaviour from a number of locations [around the world] and are keen to pursue and expand upon those reports and to hear of new reports,' says Gosford.

No one has so far video-recorded this avian incendiary behaviour. Even if the researchers catch raptors red-handed, proving intent is another issue.

'The question is,' says Quader, 'are the firehawks "selectively" picking up smouldering/burning sticks (i.e., over other items on the ground) and then "selectively" dropping them on unburnt and combustible vegetation away from the fire (i.e., over other places they could have dropped them). Personally, I think it is quite likely the birds are intentionally spreading fire – but in terms of evidential rigour, we are not quite there yet. This requires many more observations, possibly even a clever, although difficult, experiment.'

The researchers say these reports raise questions about the history and evolution of fire in the Australian savannas. They also plan to pursue another line of investigation. They have recorded, but not published, accounts narrated by Aboriginal peoples of raptors taking sticks from domestic cooking fires while camping. 'We will also seek to gather further accounts and observations about the connections between Australian Aboriginal indigenous ecological knowledge and traditional cultural practices,' says Gosford.

Aboriginal communities may have started fire-stick farming, but the birds appear to be evolving their own version of 'fire-stick hunting'.

50

Devoted Snake Moms[1]

MOST mother snakes take their jobs lightly. They find a safe refuge, lay their eggs, and leave. If they give birth to live young, they vanish as soon as the last little one pops out. Newborns face the dangers of the big, bad world on their own. Of the approximately 3,400 species of snakes, not even 3 per cent show maternal care.

The king cobra is the only species to make a nest and guard its eggs. The Indian rock python coils the full length of its muscular body around its mass of eggs. Throughout incubation, it shivers to keep them warm. But neither species stays with its young and protects them from the vagaries of life. One species is exceptional for going the extra mile in caring for its brood – the southern African python (*Python natalensis*).

Graham Alexander, of the University of Witwatersrand, South Africa, spent seven years tracking thirty-seven male and female southern African pythons at the Dinokeng Game Reserve, about 40 kilometres north of Pretoria.

The arrival of winter chill in the southern hemisphere in June signals the start of their breeding season. Females and their suitors travel a kilometre or two from their regular haunts to areas riddled with burrows.

Finding subjects for the study wasn't as difficult as single-handedly catching them. Southern African pythons are the continent's largest snakes, growing up to five metres in length and 60 kilograms in weight. The strong ones slipped through the researcher's hands and slithered down holes even as he struggled to hang on to their tails. He overcame a few and surgically inserted gadgets for tracking them and measuring their body temperature. These snakes led Alexander to others of their kind.

Driving a 200 kilometre round trip every other day from his university in Johannesburg added to his already punishing schedule. Not only did he have courses to teach, but Alexander was a single parent. Sometimes he took his three young children with him to the game reserve. 'I look back now and wonder how much danger I exposed my five-year-old daughter to,' he says.

To see how far the snakes with transmitters moved, he had to locate each one and mark its GPS coordinates. He was so pressed for time that after he noted the details of one snake, he ran to do the next one. On three occasions, hungry pythons lying in ambush mistook him for an antelope and bit him. But the non-venomous snakes realized their mistake and let him go. One python, however, started constricting his legs before

Alexander disentangled himself. His children weren't along on these trips to witness these mishaps.

At the start of the breeding season, two or three males hung around, often resting while enveloped within every would-be mamma's massive coils. The researcher observed these females had turned from their standard cryptic brown and tan colours to a stark black. Typical of species where females are the larger sex, male snakes aren't aggressive to each other as they compete for the attention of the opposite sex. One resident landowner reported an entourage of thirteen courting males following a female.

The females lazed in the sunshine at the entrance of burrows, but fled underground at any disturbance. Their all-black maternity coloration perhaps aids in the better absorption of the sun's rays but also makes them visible to any passing animals. But the males stood their ground unless a human approached too close. Maybe they thought their brown camouflage blotches made them invisible.

Three months later, the males lost interest in sex and went their way, leaving the females alone to handle parenthood. In other snake species, relationships don't last this long. Alexander doesn't know why these male pythons spent all that time with females. Since every male had at least one rival, they may have stayed to make sure they got an equal chance of fathering the next generation, a behaviour called 'scramble mating'.

The females laid their eggs inside burrows made by aardvarks, warthogs and porcupines. Alexander wasn't certain if any altercation took place underground, but the mammals continued to use their subterranean chambers even after the reptilian intruders moved in. He speculated the burrow owner and tenant used different hollows, and the parties gave each other a wide berth.

Alexander didn't count how many eggs each python laid until after the hatchlings and their mother vacated the chambers. However, he monitored the goings-on inside the nesting cavity with an infrared video camera. When he broke in after their departure, he found the smallest batch had fifteen eggs and the largest seventy-four.

Although the mothers wrapped around their eggs, they didn't rhythmically contract their muscles to raise their body temperature as Indian rock pythons do. This finding challenged accepted wisdom. Reptile experts had assumed other python species incubated by shivering. Instead, twice a day, the southern African pythons lay at the burrow entrance until they baked to around 40°C before returning underground to coil around their eggs and transfer the heat from their bodies. For the three-month incubation term, these massive mammas lived an austere lifestyle, alternately sunbathing and curling around their precious loads. They didn't eat or drink.

In December, the warmest time of year in South Africa, the young hatch. The baby snakes slit their leathery eggshells and poke their heads out, breathing air for two days before emerging. In other egg-laying species, hatching is the cue for mothers to leave and break their long fast.

Female Indian rock pythons don't wait until then, abandoning their maternal responsibilities twelve days before the due date.[2] They have done their duty, and now it is time to get some sustenance.

But the emaciated southern African python moms stayed put even after their young hatched. The little ones basked in a dense mass during the day at the burrow entrance with their mothers. The adults that had been shy and retiring earlier became bold and defensive when Alexander approached. They bunched their coils and prepared to strike, a sign they might confront any predator. At night, the hatchlings rested underground within their mothers' embrace and absorbed heat from their large bodies.

After two weeks of this routine, the young snakes sloughed their skins all at once and left the safe confines of their nests. They were ready to face the world. Only then did the thin mothers crawl away to find a meal. They hunkered down along game trails, waiting in ambush for a warm-blooded animal.

Time was of the essence because winter was around the corner. During more than six months of starvation, they lost as much as 40 per cent of their body mass. By toasting themselves every day, the mothers burned through their body's reserves. The high body temperatures revved up metabolism. Data from temperature loggers showed mother pythons raised their body heat more than 5°C higher than other pythons.

Each effort at procreation drains female southern African pythons, and they may take several years to regain peak condition. Alexander writes, 'fewer than half the adult females bred in any given year'. Some of them may wait to eat until it is too late. One scrawny female that couldn't find prey in time died. They traded their well-being to give their young a head start in life.

Why do they go to such extraordinary lengths and endanger their own lives? Newborns cannot crawl well when their bellies are full of yolk. At this vulnerable stage, the mothers possibly shield them from predators such as mongooses, meerkats and ratels that would make short work of the youngsters.

By resting on their mothers' warm coils, the newborn snakes may digest their protein stores faster.

Harry Greene, an evolutionary biologist at Cornell University, and his colleagues wrote of live-bearing maternal rattlesnakes[3] in the book *Biology of the Vipers*. The authors describe mothers staying with their young for a few days and even chasing off the predatory black racer snake from their enclosure in the lab. But these snakes didn't go to the extreme lengths that southern African pythons did.

'As a bedrock natural historian,' says Greene, 'of course I love it [Alexander's paper]. We have been vastly underestimating the behavioural complexity of snakes!'

Alexander's investigation is the first time a scientist has discovered an egg-laying snake extending her vigil to protect her offspring. In future research, he hopes to test whether females become dark to warm up faster. They didn't revert to their normal colours until long after their maternal duties ended. No other snake changes its appearance so dramatically for reproduction. He also wants to investigate whether all the babies in a clutch have the same father. A litter of puff adder hatchlings, for instance, averages about three fathers, with one having as many as six dads.[4] He has a hunch that southern African pythons may be similar.

Seven years is a long time to track snakes. Did Alexander have an inkling of this unique behaviour before he began his study? When a farm manager first told him of a mother python with babies, the researcher had dismissed that observation thinking the man had probably seen a large female with her smaller-sized suitors. A few other similar anecdotes convinced him there was more going on. Little did he realize the species would set a record for serpentine motherhood.

Despite being a dedicated dad who also wore himself out running after his pythons, Alexander didn't lose any weight. He still needs to go to the gym every day.

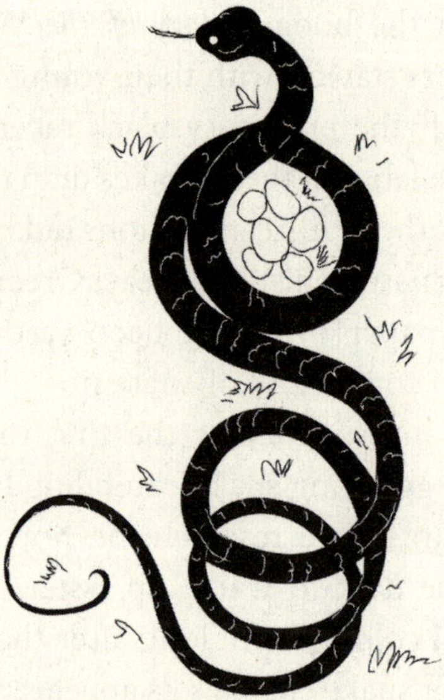

References

1. Good Singers Make the Best Dads

1 Bartsch, C., M. Weiss, and S. Kipper (2015). 'Multiple song features are related to paternal effort in common nightingales', *BMC Evolutionary Biology*, Vol. 15, No. 1, p.115. https://doi.org/10.1186/s12862-015-0390-5.

2 Kiefer, S., A. Spiess, S. Kipper, R. Mundry, C. Sommer, H. Hultsch, and D. Todt (2006). 'First-year common nightingales (Luscinia megarhynchos) have smaller songtype repertoire sizes than older males', *Ethology*, Vol. 112, No. 12, pp. 1217–24. https://doi.org/10.1111/j.1439-0310.2006.01283.x

3 Roth, T., P. Sprau, R. Schmidt, M. Naguib, and V. Amrhein (2009). 'Sex-specific timing of mate searching and territory prospecting in the nightingale: nocturnal life of females', *Proceedings of the Royal*

Society of London B: Biological Sciences, Vol. 276, Issue 1664. https://doi.org/10.1098/rspb.2008.1726

4 Amrhein, V., H.P. Kunc, and M. Naguib (2004). 'Non–territorial nightingales prospect territories during the dawn chorus', *Proceedings of the Royal Society of London B: Biological Sciences,* Vol. 271, Suppl. 4, pp. S167–S169. https://doi.org/10.1098/rsbl.2003.0133

2. Chimpanzees in Mourning

1 Van Leeuwen, E.J., I.C. Mulenga, M.D. Bodamer, and K.A. Cronin (2016). 'Chimpanzees' responses to the dead body of a 9-year-old group member', *American Journal of Primatology*, Vol. 78, No. 9, pp. 914–22. https://doi.org/10.1002/ajp.22560

2 Fashing, P.J., N. Nguyen, T.S. Barry, C.B. Goodale, R.J. Burke, S.C. Jones, J.T. Kerby, L.M. Lee, N.O. Nurmi, and V.V. Venkataraman (2011). 'Death among geladas (Theropithecus gelada): a broader perspective on mummified infants and primate thanatology', *American Journal of Primatology*, Vol. 73, No. 5, pp. 405–09. https://doi.org/10.1002/ajp.20902

3 Van Leeuwen, E.J., K.A. Cronin, and D.B. Haun (2017), 'Tool use for corpse cleaning in chimpanzees', *Scientific Reports*, 7, Article No. 44091. https://doi.org/10.1038/srep44091

4 Gonçalves, A., and S. Carvalho (2019). 'Death among primates: a critical review of non-human primate interactions towards their dead and dying', *Biological Reviews*, https://doi.org/10.1111/brv.12512

5 Boesch, C. (1991). 'The effects of leopard predation on grouping patterns in forest chimpanzees', *Behaviour,* Vol. 117, Issue 3–4, pp. 220–42. https://www.jstor.org/stable/4534940

6 Anderson J.R. (2018) 'Chimpanzees and death', Philosophical Transactions of the Royal Society B 373. https://doi.org/10.1098/rstb.2017.0257

7 Anderson, J.R., A. Gillies, and L.C. Lock (2010). 'Pan thanatology', *Current Biology*, Vol. 20, No. 8, R349–R351. https://doi.org/10.1016/j.cub.2010.02.010

8 Goodall, J. (2010). *Through a window: my thirty years with the chimpanzees of Gombe*, Boston, Massachusetts: Houghton Mifflin Harcourt

3. The Eye of the Chameleon

1 Katz, H.K., A. Lustig, T. Lev-Ari, Y. Nov, E. Rivlin, and G. Katzir (2015). 'Eye movements in chameleons are not truly independent – evidence from simultaneous monocular tracking of two targets', *Journal of Experimental Biology*, Vol. 218, No. 13, pp. 2097–2105. https://doi.org/10.1242/jeb.113084

2 Ott, M., and F. Schaeffel (1995). 'A negatively powered lens in the chameleon', *Nature*, Vol. 373, No. 6516, p. 692. https://doi.org/10.1038/373692a0

3 Harkness, L. (1977). 'Chameleons use accommodation cues to judge distance', *Nature*, Vol. 267, No. 5609, p. 346. https://doi.org/10.1038/267346a0

4. The Oddest Bird in the World

1 Le Duc, D., G. Renaud, A. Krishnan, M.S. Almén, L. Huynen, S.J. Prohaska, M. Ongyerth, B.D. Bitarello, H.B. Schiöth, M. Hofreiter, and P.F. Stadler (2015). 'Kiwi genome provides insights into evolution of a nocturnal lifestyle', *Genome Biology*, Vol. 16, No. 1, p. 147. https://doi.org/10.1186/s13059-015-0711-4

2 Cunningham, S., I. Castro, and M. Alley (2007). 'A new prey-detection mechanism for kiwi (Apteryx spp.) suggests convergent evolution between paleognathous and neognathous birds', *Journal of Anatomy*, Vol. 211, No. 4, pp. 493–502. https://doi.org/10.1111/j.1469-7580.2007.00786.x

3 Castro, I., S.J. Cunningham, A.C. Gsell, K. Jaffe, A. Cabrera, and C. Liendo (2010). 'Olfaction in birds: a closer look at the kiwi (Apterygidae)', *Journal of Avian Biology*, Vol. 41, No. 3, pp. 213–18. https://doi.org/10.1111/j.1600-048X.2010.05010.x

4 Martin, G. R., K.J. Wilson, J.M. Wild, S. Parsons, M.F. Kubke, and J. Corfield (2007). 'Kiwi forego vision in the guidance of their

nocturnal activities', *Plos One*, Vol. 2, No. 2, e198. https://doi.org/10.1371/journal.pone.0000198

5 Martin, G.R. (2017). *The Sensory Ecology of birds*. Oxford, UK: Oxford University Press.

6 Calder III, W.A. (1979). 'The kiwi and egg design: evolution as a package deal', *Bioscience*, Vol. 29, No. 8, pp. 461–67. https://doi.org/10.2307/1307538

5. Vaccinated Bees

1 Salmela, H., G.V. Amdam, and D. Freitak (2015). 'Transfer of immunity from mother to offspring is mediated via egg-yolk protein vitellogenin', *PLoS Pathogens*, Vol. 11, No. 7, e1005015. https://doi.org/10.1371/journal.ppat.1005015

2 Krautz R., B. Arefin, and U. Theopold (2014). 'Damage signals in the insect immune response', *Frontiers in Plant Science*, 2014;5:342. https://doi.org/10.3389/fpls.2014.00342.

3 Lemaitre, B., and J. Hoffmann (2007). 'The Host Defense of Drosophila melanogaster', *Annual Review of Immunology*, Vol. 25, No. 1, pp. 697–743. https://doi.org/10.1146/annurev.immunol.25.022106.141615

4 Li, H., and S. Zhang (2017). 'Functions of Vitellogenin in Eggs', in Kloc M. (ed.), *Oocytes. Results and Problems in Cell Differentiation*, Vol 63. Springer, Cham https://doi.org/10.1007/978-3-319-60855-6_17

6. The Scent of a Fur Seal

1 Stoffel, M.A., B.A. Caspers, J. Forcada, A. Giannakara, M. Baier, L. Eberhart-Phillips, C. Müller, and J.I. Hoffman (2015). 'Chemical fingerprints encode mother–offspring similarity, colony membership, relatedness, and genetic quality in fur seals', *Proceedings of the National Academy of Sciences*, Vol. 112, No. 36, E5005–E5012. https://doi.org/10.1073/pnas.1506076112

2 Hoffman, J. I., and W. Amos (2005). 'Does kin selection influence fostering behaviour in Antarctic fur seals (Arctocephalus gazella)?', *Proceedings of the Royal Society B: Biological Sciences*, Vol. 272, No. 1576, pp. 2017–22. https://doi.org/10.1098/rspb.2005.3176

3 Dobson, F.S., and P. Jouventin (2003). 'How mothers find their pups in a colony of Antarctic fur seals', *Behavioural Processes*, Vol. 61, Nos 1&2, pp. 77-85. https://doi.org/10.1016/S0376-6357(02)00164-X

4 Charrier, I., N. Mathevon, and P. Jouventin (2002). 'How does a fur seal mother recognize the voice of her pup? An experimental study of Arctocephalus tropicalis', *Journal of Experimental Biology*, Vol. 205, No. 5, pp. 603–12.

5 Aubin, T., P. Jouventin, and I. Charrier (2015). 'Mother Vocal Recognition in Antarctic Fur Seal Arctocephalus gazella Pups: a two-step process', *PloS One*, Vol. 10, No. 9, e0134513. https://doi.org/10.1371/journal.pone.0134513

6 Pitcher, B.J., R.G. Harcourt, B. Schaal, and I. Charrier (2011). 'Social olfaction in marine mammals: wild female Australian sea lions can identify their pup's scent', *Biology Letters*, Vol. 7, No. 1, pp. 60–62. https://doi.org/10.1098/rsbl.2010.0569

7 Dobson, F.S., and P. Jouventin (2003). 'How mothers find their pups in a colony of Antarctic fur seals', *Behavioural Processes*, Vol. 61, Nos 1&2, pp. 77-85. https://doi.org/10.1016/S0376-6357(02)00164-X

8 Varendi, H., and R.H. Porter (2001). 'Breast odour as the only maternal stimulus elicits crawling towards the odour source', *Acta Paediatrica*, Vol. 90, No. 4, pp. 372–75. https://doi.org/10.1111/j.1651-2227.2001.tb00434.x

9 Eidelman, A., M. Kaitz, A. Good, and A. Rokem (1987). 'Mothers' Recognition of Their Newborns by Olfactory Cues', *Developmental Psycobiology*, Vol. 20, No. 6, 180A. https://doi.org/10.1002/dev.420200604

10 Gustafsson, E., F. Levréro, D. Reby, and N. Mathevon (2013). 'Fathers are just as good as mothers at recognizing the cries of their baby', *Nature Communications*, Vol. 4, Article No. 1698. https://doi.org/10.1038/ncomms2713

7. Conservative Chimpanzees

1 Koops, K., C. Schöning, M. Isaji, and C. Hashimoto (2015). 'Cultural differences in ant-dipping tool length between neighbouring

chimpanzee communities at Kalinzu, Uganda', *Scientific Reports*, 5, Article No. 12456. https://doi.org/10.1038/srep12456

8. Sex Change in Bearded Dragons

1 Holleley, C.E., D. O'Meally, S.D. Sarre, J.A.M. Graves, T. Ezaz, K. Matsubara, B. Azad, X. Zhang, and A. Georges (2015). 'Sex reversal triggers the rapid transition from genetic to temperature-dependent sex', *Nature*, Vol. 523, No. 7558, p. 79. https://doi.org/10.1038/nature14574

2 Quinn, A.E., A. Georges, S.D. Sarre, F. Guarino, T. Ezaz, and J.A.M. Graves (2007). 'Temperature sex reversal implies sex gene dosage in a reptile', *Science*, Vol. 316, No. 5823, p. 411. https://doi.org/10.1126/science.1135925

9. Wasps Enslave Spiders to Weave for Them

1 Takasuka, K., T. Yasui, T. Ishigami, K. Nakata, R. Matsumoto, K. Ikeda, and K. Maeto (2015). 'Host manipulation by an ichneumonid spider ectoparasitoid that takes advantage of preprogrammed web-building behaviour for its cocoon protection', *Journal of Experimental Biology*, Vol. 218, No. 15, pp. 2326–32. https://doi.org/10.1242/jeb.122739

10. Making Peace with Wolves

1 Venkataraman, V.V., J.T. Kerby, N. Nguyen, Z.T. Ashenafi, and P.J. Fashing (2015). 'Solitary Ethiopian wolves increase predation success on rodents when among grazing gelada monkey herds', *Journal of Mammalogy*, Vol. 96, No. 1, pp. 129–37. https://doi.org/10.1093/jmammal/gyu013

11. The Amorous Octopus

1 Caldwell, R.L., R. Ross, A. Rodaniche, and C.L. Huffard (2015). 'Behavior and Body Patterns of the Larger Pacific Striped Octopus',

PloS One, Vol. 10, Issue 8, e0134152. https://doi.org/10.1371/journal.pone.0134152

12. Pregnant Fathers

1 Whittington, C.M., O.W. Griffith, W. Qi, M.B. Thompson, and A.B. Wilson (2015). 'Seahorse brood pouch transcriptome reveals common genes associated with vertebrate pregnancy', *Molecular Biology and Evolution*, Vol. 32, No. 12, pp. 3114–31. https://doi.org/10.1093/molbev/msv177

13. Why Don't Elephants Get Cancer?

1 Abegglen, L.M., A.F. Caulin, A. Chan, K. Lee, R. Robinson, M.S. Campbell, W.K. Kiso, D.L. Schmitt, P.J. Waddell, S. Bhaskara, and S.T. Jensen (2015). 'Potential mechanisms for cancer resistance in elephants and comparative cellular response to DNA damage in humans', *Jama*, Vol. 314, No. 17, pp.1850–60. https://doi.org/10.1001/jama.2015.13134

2 Dikshit, R., P.C. Gupta, C. Ramasundarahettige, V. Gajalakshmi, L. Aleksandrowicz, R. Badwe, R. Kumar, S. Roy, W. Suraweera, F. Bray, and M. Mallath (2012). 'Cancer mortality in India: a nationally representative survey', *The Lancet*, Vol. 379, No. 9828, pp.1807–16. https://doi.org/10.1016/S0140-6736(12)60358-4

3 Anand, P., A.B. Kunnumakara, C. Sundaram, K.B. Harikumar, S.T. Tharakan, O.S. Lai, B. Sung, and B.B. Aggarwal (2008). 'Cancer is a preventable disease that requires major lifestyle changes', *Pharmaceutical Research*, Vol. 25, No. 9, pp.2097–116. https://doi.org/10.1007/s11095-008-9661–9

4 Peto, R., F.J. Roe, P.N. Lee, L. Levy, and J. Clack (1975). 'Cancer and ageing in mice and men', *British Journal of Cancer*, No. 32, pp. 411–26. https://doi.org/10.1038/bjc.1975.242

5 Lane, D.P. (1992). 'Cancer: p53, guardian of the genome', *Nature*, Vol. 358, No. 6381, pp. 15–16. https://doi.org/10.1038/358015a0

6 Sulak, M., L. Fong, K. Mika, S. Chigurupati, L. Yon, N.P. Mongan, R.D. Emes, and V.J. Lynch, (2016). 'TP53 copy number expansion

is associated with the evolution of increased body size and an enhanced DNA damage response in elephants', *Elife*, No. 5, p. e11994. https://doi.org/10.7554/eLife.11994

7 Vazquez, J.M., M. Sulak, S. Chigurupati, and V.J. Lynch (2018). 'A zombie LIF gene in elephants is upregulated by TP53 to induce apoptosis in response to DNA damage', *Cell Reports*, Vol. 24, Issue 7, pp. 1765–76. https://doi.org/10.1016/j.celrep.2018.07.042

8 Aghajanova, L. (2004). 'Leukemia inhibitory factor and human embryo implantation', *Annals of the New York Academy of Sciences*, Vol. 1034, Issue 1, pp. 176–83. https://doi.org/10.1196/annals.1335.020

9 Caulin, A.F. ,and C.C. Maley (2011). 'Peto's Paradox: evolution's prescription for cancer prevention', *Trends in Ecology & Evolution*, Vol. 26, Issue 4, pp. 175–82. https://doi.org/10.1016/j.tree.2011.01.002

10 Seluanov, A., C. Hine, J. Azpurua, M. Feigenson, M. Bozzella, Z. Mao, K.C. Catania, and V. Gorbunova (2009). 'Hypersensitivity to contact inhibition provides a clue to cancer resistance of naked mole-rat', *Proceedings of the National Academy of Sciences*, Vol. 106, No. 46, pp.19352–57. https://doi.org/10.1073/pnas.0905252106

11 Tian, X., J. Azpurua, C. Hine, A. Vaidya, M. Myakishev-Rempel, J. Ablaeva, Z. Mao, E. Nevo, V. Gorbunova. and A. Seluanov (2013). 'High-molecular-mass hyaluronan mediates the cancer resistance of the naked mole rat', *Nature*, No. 499, pp. 346–49. https://doi.org/10.1038/nature12234

14. Sticklebacks Hold Their Water

1 Madsen, S.S., C. Weber, A.M. Nielsen, M. Mohiseni, M.C. Bosssus, C.K. Tipsmark, and B. Borg (2015). 'Sexual maturation and changes in water and salt transport components in the kidney and intestine of three-spined stickleback (Gasterosteus aculeatus L.)', *Comparative Biochemistry and Physiology Part A: Molecular & Integrative Physiology*, Vol. 188, pp. 107–19. https://doi.org/10.1016/j.cbpa.2015.06.021

2 Katsiadaki, I., A.P. Scott, and I. Mayer (2002). 'The potential of the three-spined stickleback (Gasterosteus aculeatus L.) as a combined

biomarker for oestrogens and androgens in European waters', *Marine Environmental Research*, Vol. 54, Issues 3-5, pp. 725–28. https://doi.org/10.1016/S0141-1136(02)00110-1

15. 'Who Am I?'

1 Louder, M.I., M.P. Ward, W.M. Schelsky, M.E. Hauber, and J.P. Hoover (2015). 'Out on their own: a test of adult-assisted dispersal in fledgling brood parasites reveals solitary departures from hosts', *Animal Behaviour*, Vol. 110, pp. 29–37. https://doi.org/10.1016/j.anbehav.2015.09.009

2 Hoover, J.P., and S.K. Robinson (2007). 'Retaliatory mafia behavior by a parasitic cowbird favors host acceptance of parasitic eggs', *Proceedings of the National Academy of Sciences*, Vol. 104, No. 11, pp. 4479–83. https://doi.org/10.1073/pnas.0609710104

3 Hahn, D.C., and R.C. Fleischer (1995). 'DNA fingerprint similarity between female and juvenile brown-headed cowbirds trapped together', *Animal Behaviour*, Vol. 49, Issue. 6, pp. 1577–80. https://doi.org/10.1016/0003-3472(95)90079-9

4 Hauber, M.E., S.A. Russo, and P.W. Sherman (2001). 'A password for species recognition in a brood-parasitic bird', *Proceedings of the Royal Society of London, Series B: Biological Sciences*, Vol. 268, Issue 1471, pp. 1041–48. https://doi.org/10.1098/rspb.2001.1617

5 Lynch, K.S., A. Gaglio, E. Tyler, J. Coculo, M.I. Louder, and M.E. Hauber (2017). 'A neural basis for password-based species recognition in an avian brood parasite', *Journal of Experimental Biology*, jeb-158600. https://doi.org/10.1242/jeb.158600

16. Rodent Monogamy

1 Okhovat, M., A. Berrio, G. Wallace, A.G. Ophir, and S.M. Phelps (2015). 'Sexual fidelity trade-offs promote regulatory variation in the prairie vole brain', *Science*, Vol. 350, Issue 6266, pp. 1371–74. https://doi: 10.1126/science.aac5791

2 Ophir, A.G., J.O. Wolff, and S.M. Phelps (2008). 'Variation in neural V1aR predicts sexual fidelity and space use among male prairie

voles in semi-natural settings', *Proceedings of the National Academy of Sciences*, Vol. 105, No. 4, pp. 1249–54. https://doi.org/10.1073/pnas.0709116105

3 Smith, K.S., A.J. Tindell, J.W. Aldridge, and K.C. Berridge (2009). 'Ventral pallidum roles in reward and motivation', *Behavioural Brain Research*, Vol. 196, Issue 2, pp. 155–67. https://doi.org/10.1016/j.bbr.2008.09.038

4 Walum, H., L. Westberg, S. Henningsson, J.M. Neiderhiser, D. Reiss, W. Igl, J.M. Ganiban, E.L. Spotts, N.L. Pedersen, E. Eriksson, and P. Lichtenstein (2008). 'Genetic variation in the vasopressin receptor 1a gene (AVPR1A) associates with pair-bonding behavior in humans', *Proceedings of the National Academy of Sciences*, Vol. 105, No. 37, pp.14153–56. https://doi.org/10.1073/pnas.0803081105

5 Wang, H., F. Duclot, Y. Liu, Z. Wang, and M. Kabbaj (2013). 'Histone deacetylase inhibitors facilitate partner preference formation in female prairie voles', *Nature Neuroscience*, 16, pp. 919–24. https://doi.org/10.1038/nn.3420

6 Lim, M.M., Z. Wang, D.E. Olazábal, X. Ren, E.F. Terwilliger, and L.J. Young (2004). 'Enhanced partner preference in a promiscuous species by manipulating the expression of a single gene', *Nature*, 429, p. 754–57. https://doi.org/10.1038/nature02539

7 Vann, S.D., J.P. Aggleton, and E.A. Maguire (2009). 'What does the retrosplenial cortex do?, *Nature Reviews Neuroscience*, 10, pp. 792–802. https://doi.org/10.1038/nrn2733

8 Ophir, A.G., J.O. Wolff, and S.M. Phelps (2008). 'Variation in neural V1aR predicts sexual fidelity and space use among male prairie voles in semi-natural settings', *Proceedings of the National Academy of Sciences*, Vol. 105, No. 4, pp. 1249–54. https://doi.org/10.1073/pnas.0709116105

9 Zietsch, B.P., L. Westberg, P. Santtila, and P. Jern (2015). 'Genetic analysis of human extrapair mating: heritability, between-sex correlation, and receptor genes for vasopressin and oxytocin', *Evolution and Human Behavior*, Vol. 36, Issue 2, pp. 130–36. https://doi.org/10.1016/j.evolhumbehav.2014.10.001

17. The Snake that Dodges Predators?

1 Miller, A.K., B. Maritz, S. McKay, X. Glaudas, and G.J. Alexander (2015). 'An ambusher's arsenal: chemical crypsis in the puff adder (*Bitis arietans*)', *Proceedings of the Royal Society B: Biological Sciences*, Vol. 282, No. 1821, pp. 2015–2182. https://doi.org/10.1098/rspb.2015.2182

18. Food or Mates?

1 Firth, J.A., B. Voelkl, D.R. Farine, and B.C. Sheldon (2015). 'Experimental evidence that social relationships determine individual foraging behavior', *Current Biology*, Vol. 25, Issue 23, pp. 3138–43. https://doi.org/10.1016/j.cub.2015.09.075

19. Dodging Old Age

1 Giraldo, Y.M., J.F. Kamhi, V. Fourcassié, M. Moreau, S.K. Robson, A. Rusakov, L. Wimberly, A. Diloreto, A. Kordek, and J.F. Traniello (2016). 'Lifespan behavioural and neural resilience in a social insect', *Proceedings of the Royal Society B: Biological Sciences*, Vol. 283, Issue 1822, 2015.2603. https://doi.org/10.1098/rspb.2015.2603

2 Rueppell, O., S. Christine, C. Mulcrone, and L. Groves (2007). 'Aging without functional senescence in honey bee workers', *Current Biology: CB*, Vol. 17, Issue 8, R274–R275. https://doi.org/10.1016/j.cub.2007.02.015

3 Behrends, A., R. Scheiner, N. Baker, and G.V. Amdam (2007). 'Cognitive aging is linked to social role in honey bees (Apis mellifera)', *Experimental Gerontology*, Vol. 42, No. 12, pp. 1146–53. https://doi.org/10.1016/j.exger.2007.09.003

20. Shark Navigation

1 Nosal, A.P., Y. Chao, J.D. Farrara, F. Chai, and P.A. Hastings (2016). 'Olfaction contributes to pelagic navigation in a coastal shark', *PloS One*, 11(1), e0143758. https://doi.org/10.1371/journal.pone.0143758

2 Gardiner, J.M., and J. Atema (2007). 'Sharks need the lateral line to locate odor sources: rheotaxis and eddy chemotaxis', *Journal of Experimental Biology*, Vol. 210, No. 11, pp. 1925–34. https://doi.org/10.1242/jeb.000075

3 Yopak, K.E., T.J. Lisney, and S.P. Collin (2015). 'Not all sharks are "swimming noses": variation in olfactory bulb size in cartilaginous fishes', *Brain Structure and Function*, 220, pp. 1127–43. https://doi.org/10.1007/s00429-014-0705-0

4 Jacobs, L.F. (2012). 'From chemotaxis to the cognitive map: the function of olfaction', *Proceedings of the National Academy of Sciences*, No. 109 (Supplement 1), pp. 10693–700. https://doi.org/10.1073/pnas.1201880109

5 Gardiner, J.M., J. Atema, R.E. Hueter, and P.J. Motta (2014). 'Multisensory integration and behavioral plasticity in sharks from different ecological niches', *PLoS One*, Vol. 9, Issue 4, e93036. https://doi.org/10.1371/journal.pone.0093036

21. A Parasite that Meddles with Behaviour?

1 Poirotte, C., P.M. Kappeler, B. Ngoubangoye, S. Bourgeois, M. Moussodji, and M.J. Charpentier (2016). 'Morbid attraction to leopard urine in Toxoplasma-infected chimpanzees', *Current Biology*, Vol. 26, Issue 3, R98–R99. https://doi.org/10.1016/j.cub.2015.12.020

2 House, P.K., A. Vyas, and R. Sapolsky (2011). 'Predator cat odors activate sexual arousal pathways in brains of Toxoplasma gondii infected rats', *Plos One*, 6 (8), e23277. https://doi.org/10.1371/journal.pone.0023277

3 Webster, J.P. (2001). 'Rats, cats, people and parasites: the impact of latent toxoplasmosis on behaviour', *Microbes and Infection*, Vol. 3, Issue 12, pp. 1037–45. https://doi.org/10.1016/S1286-4579(01)01459-9

4 Webster, J.P. (2001). 'Rats, cats, people and parasites: the impact of latent toxoplasmosis on behaviour. *Microbes and Infection*,

Vol. 3, Issue 12, pp. 1037–45. https://doi.org/10.1016/S1286-4579(01)01459-9

5 Singh, S., A. Munawwar, S. Rao, S. Mehta, and N.K. Hazarika (2014). 'Serologic prevalence of Toxoplasma gondii in Indian women of child bearing age and effects of social and environmental factors', *PLoS Neglected Tropical Diseases*, Vol. 8, No. 3, e2737. https://doi.org/10.1371/journal.pntd.0002737

6 Flegr, J., S. Zitková, P Kodym, and D. Frynta (1996). 'Induction of changes in human behaviour by the parasitic protozoan Toxoplasma gondii', *Parasitology*, Vol. 113, Issue 1, pp. 49–54. https://doi.org/10.1017/S0031182000066269

7 Flegr, J. (2013). 'How and why Toxoplasma makes us crazy', *Trends in Parasitology*, Vol. 29, Issue 4, pp. 156–63. https://doi.org/10.1016/j.pt.2013.01.007

8 Berenreiterová, M., J. Flegr, A.A. Kuběna, and P. Němec (2011). 'The distribution of Toxoplasma gondii cysts in the brain of a mouse with latent toxoplasmosis: implications for the behavioral manipulation hypothesis', *PloS One*, Vol. 6, No. 12, e28925. https://doi.org/10.1371/journal.pone.0028925

9 Boesch, C. (1991). 'The effects of leopard predation on grouping patterns in forest chimpanzees', *Behaviour*, Vol. 117, Issue 3-4, pp. 220–241. https://doi.org/10.1163/156853991X00544

10 Zuberbühler, K., and D. Jenny, (2002). 'Leopard predation and primate evolution', *Journal of Human Evolution*, Vol. 43, Issue 6, pp. 873–86. https://doi.org/10.1006/jhev.2002.0605

11 Tsukahara, T. (1993). 'Lions eat chimpanzees: the first evidence of predation by lions on wild chimpanzees', *American Journal of Primatology*, Vol. 29, Issue 1, pp. 1–11. https://doi.org/10.1002/ajp.1350290102

12 Boillat, M., P-M. Hammoudi, S.K. Dogga, S. Pages, M. Goubran, I. Rodriguez, and D. Soldati-Favre (2020). Neuroinflammation-associated aspecific manipulation of mouse predator fear by *Toxoplasma gondii. Cell Reports,* Vol. 30, Issue 2, pp. 320-334. https://doi.org/10.1016/j.celrep.2019.12.019

22. High Speed Bird Vision

1 Boström, J.E., M. Dimitrova, C. Canton, O. Håstad, A. Qvarnström, and A. Ödeen (2016). 'Ultra-rapid vision in birds', *Plos One*, Vol. 11, No. 3, e0151099. https://doi.org/10.1371/journal.pone.0151099

2 Healy, K., L. McNally, G.D. Ruxton, N. Cooper, and A.L. Jackson (2013). 'Metabolic rate and body size are linked with perception of temporal information', *Animal Behaviour*, Vol. 86, Issue 4, pp. 685–96. https://doi.org/10.1016/j.anbehav.2013.06.018

3 Lisney, T.J., D. Rubene, J. Rózsa, H. Løvlie, O. Håstad, and A. Ödeen (2011). 'Behavioural assessment of flicker fusion frequency in chicken Gallus gallus domesticus', *Vision Research*, Vol. 51, Issue 12, pp.1324–32. https://doi.org/10.1016/j.visres.2011.04.009

4 Lisney, T.J., B. Ekesten, R. Tauson, O. Håstad, and A. Ödeen (2012). 'Using electroretinograms to assess flicker fusion frequency in domestic hens Gallus gallus domesticus', *Vision Research*, No. 62, pp. 125–33. https://doi.org/10.1016/j.visres.2012.04.002

5 Boström, J.E., N.K. Haller, M. Dimitrova, A. Ödeen, and A. Kelber (2017). 'The flicker fusion frequency of budgerigars (Melopsittacus undulatus) revisited', *Journal of Comparative Physiology A*, 203, pp. 15–22. https://doi.org/10.1007/s00359-016-1130-z

23. Empathetic Rodents

1 Burkett, J.P., E. Andari, Z.V. Johnson, D.C. Curry, F.B. de Waal, and L.J. Young (2016). 'Oxytocin-dependent consolation behavior in rodents', *Science*, Vol. 351, No. 6271, pp. 375–78. https://doi.org/10.1126/science.aac4785

2 De Waal, F.B., and A. van Roosmalen (1979). 'Reconciliation and consolation among chimpanzees', *Behavioral Ecology and Sociobiology*,5, pp. 55–66. https://doi.org/10.1007/BF00302695

3 Romero, T., M.A. Castellanos, and F.B. de Waal (2010). 'Consolation as possible expression of sympathetic concern among chimpanzees', *Proceedings of the National Academy of Sciences*, Vol. 107, No. 27, pp.12110–15. https://doi.org/10.1073/pnas.1006991107

4 Sato, N., L. Tan, K. Tate, and M. Okada (2015). 'Rats demonstrate helping behavior toward a soaked conspecific', *Animal Cognition*, 18, pp.1039–47. https://doi.org/10.1007/s10071-015-0872-2

5 Domes, G., M. Heinrichs, A. Michel, C. Berger, and S.C. Herpertz (2007). 'Oxytocin improves "mind-reading" in humans', *Biological Psychiatry*, Vol. 61, Issue 6, pp. 731–33. https://doi.org/10.1016/j.biopsych.2006.07.015

6 Gu, X., Z. Gao, X. Wang, X. Liu, R.T. Knight, P.R. Hof, and J. Fan (2012). 'Anterior insular cortex is necessary for empathetic pain perception', *Brain*, Vol. 135, Issue 9, pp. 2726–35. https://doi.org/10.1093/brain/aws199

7 Andari, E., J.R. Duhamel, T. Zalla, E. Herbrecht, M. Leboyer, and A. Sirigu (2010). 'Promoting social behavior with oxytocin in high-functioning autism spectrum disorders', *Proceedings of the National Academy of Sciences*, Vol. 107, No. 9, pp. 4389–94. https://doi.org/10.1073/pnas.0910249107

24. Bizarre Spider Sex

1 Gregorič, M., K. Šuen, R.C. Cheng, S. Kralj-Fišer, and M. Kuntner (2016). 'Spider behaviors include oral sexual encounters', *Scientific Reports*, No. 6, Article No. 25128. https://doi.org//10.1038/srep25128

2 Kuntner, M., and I. Agnarsson (2010). 'Web gigantism in Darwin's bark spider, a new species from Madagascar (*Araneidae: Caerostris*)', *The Journal of Arachnology*, Vol. 38, No. 2, pp. 346–57. https://doi.org/10.1636/B09-113.1

3 Blackledge, T., M. Kuntner, and I. Agnarsson (2010). 'Bioprospecting finds the toughest biological material: extraordinary silk from a giant riverine orb spider', *Plos One*, Vol. 5, No. 9. https://doi.org/10.1371/journal.pone.0011234

4 Lee, Q.Q., J. Oh, S. Kralj-Fišer, M. Kuntner, and D. Li, (2012). 'Emasculation: gloves-off strategy enhances eunuch spider endurance', *Biology Letters*, Vol. 8, Issue 5, pp. 733–35. https://doi.org/10.1098/rsbl.2012.0285

25. How Do Echidnas Survive Wildfires?

1 Nowack, J., C.E. Cooper, and F. Geiser (2016). 'Cool echidnas survive the fire', *Proceedings of the Royal Society B: Biological Sciences*, Vol. 283, Issue 1828. https://doi.org/10.1098/rspb.2016.0382

2 Ruf, T., and F. Geiser (2015). 'Daily torpor and hibernation in birds and mammals', *Biological Reviews*, Vol. 90, Issue 3, pp. 891–926. https://doi.org/10.1111/brv.12137

3 Stawski, C., J.K. Matthews, G. Körtner, and F. Geiser (2015). 'Physiological and behavioural responses of a small heterothermic mammal to fire stimuli', *Physiology & Behavior*, No. 151, pp. 617–22. https://doi.org/10.1016/j.physbeh.2015.09.002

4 Brice, P.H., G.C. Grigg, L.A. Beard, and J.A. Donovan (2002). 'Patterns of activity and inactivity in echidnas (*Tachyglossus aculeatus*) free-ranging in a hot dry climate: correlates with ambient temperature, time of day and season', *Australian Journal of Zoology*, Vol. 50, No. 5, pp. 461–75. https://doi.org/10.1071/ZO01080

5 Geiser, F., and G. Körtner (2010). 'Hibernation and daily torpor in Australian mammals', *Australian Zoologist*, Vol. 35, No. 2, pp. 204–15. https://doi.org/10.7882/AZ.2010.009

6 Nowack, J., A.D. Rojas, G. Körtner, and F. Geiser (2015). 'Snoozing through the storm: torpor use during a natural disaster', *Scientific Reports*, No. 5, Article No. 11243. https://doi.org/10.1038/srep11243

7 Lovegrove, B.G., K.D. Lobban, and D.L. Levesque (2014). 'Mammal survival at the Cretaceous–Palaeogene boundary: metabolic homeostasis in prolonged tropical hibernation in tenrecs', *Proceedings of the Royal Society B: Biological Sciences*, Vol. 281, Issue 1796, 2014.1304. https://doi.org/10.1098/rspb.2014.1304

8 Geiser, F., C. Stawski, C.B. Wacker, and J Nowack (2017). 'Phoenix from the Ashes: Fire, Torpor, and the Evolution of Mammalian Endothermy', *Frontiers in Physiology*, No. 8, p. 842. https://doi.org/10.3389/fphys.2017.00842

9 Geiser, F., and C. Turbill (2009). 'Hibernation and daily torpor minimize mammalian extinctions', *Naturwissenschaften*, Vol. 96, No. 10, pp. 1235–40. https://doi.org/10.1007/s00114-009-0583-0

26. Plants with Private Armies

1 Lortzing, T., O.W. Calf, M. Boehlke, J. Schwachtje, J. Kopka, D. Geuß, S. Kosanke, N.M. van Dam, and A. Steppuhn (2016). 'Extrafloral nectar secretion from wounds of Solanum dulcamara', *Nature Plants*, Vol. 2, No. 5. 16056. https://doi.org/10.1038/nplants.2016.56

2 Staab, M., F. Fornoff, A.M. Klein, and N. Blüthgen (2017). 'Ants at plant wounds: a little-known trophic interaction with evolutionary implications for ant-plant interactions', *The American Naturalist*, Vol. 190, No. 3, pp. 442–50. http://doi.org/10.1086/692735

27. Did Moby Dick Sink the Pequod?

1 Panagiotopoulou, O., P. Spyridis, H.M. Abraha, D.R. Carrier, and T.C. Pataky (2016). 'Architecture of the sperm whale forehead facilitates ramming combat', *PeerJ*, 4, e1895. https://doi.org/10.7717/peerj.1895

2 Baird, S.F. (1878). 'Report of the Commissioner of Fish and Fisheries for 1875-1876', United States Commission for Fish and Fisheries.

3 Carrier, D.R., S.M. Deban, and J. Otterstrom (2002). 'The face that sank the Essex: potential function of the spermaceti organ in aggression', *Journal of Experimental Biology*, Vol. 205, pp. 1755–63. https://jeb.biologists.org/content/205/12/1755

4 Gowans, S., and L. Rendell (1999). 'Head-Butting in Northern Bottlenose Whales (*Hyperoodon Ampullatus*): a possible function for big heads?' *Marine Mammal Science*, Vol. 15, Issue 4, pp.1342–50. https://doi.org/10.1111/j.1748-7692.1999.tb00896.x

5 Fulling, G.L., T.A. Jefferson, D. Fertl, J.C.S. Vega, C.S. Oedekoven, and S.A. Kuczaj (2017). 'Sperm Whale (*Physeter macrocephalus*) collision with a research vessel: accidental collision or deliberate

ramming?' *Aquatic Mammals*, Vol. 43, No. 4, pp. 421–29. https://doi.org/10.1578/AM.43.4.2017.421

28. The Giraffe's Neck Solution

1 Agaba, M., E. Ishengoma, W.C. Miller, B.C. McGrath, C.N. Hudson, O.C.B. Reina, A. Ratan, R. Burhans, R. Chikhi, P. Medvedev, and C.A. Praul (2016). 'Giraffe genome sequence reveals clues to its unique morphology and physiology', *Nature Communications*, No. 7, Article No.11519. https://doi.org/10.1038/ncomms11519

2 Hargens, A.R., R.W. Millard, K. Pettersson, and K. Johansen (1987). 'Gravitational haemodynamics and oedema prevention in the giraffe', *Nature*, Vol. 329, No. 6134, pp. 59–60. https://doi.org/10.1038/329059a0

3 Mitchell, G., J.P. Bobbitt, and S. Devries (2008). 'Cerebral perfusion pressure in giraffe: modelling the effects of head-raising and-lowering', *Journal Of Theoretical Biology*, Vol. 252, Issue 1, pp. 98–108. https://doi.org/10.1016/j.jtbi.2008.01.017

4 Frackowiak, H., and H. Jakubowski (2008). 'Arterial vascularization of the giraffe brain', *Ann. Zool. Fennici*, No. 45, pp. 353–59. http://www.annzool.net/PDF/anz45-free/anz45-353.pdf

5 Mitchell, G., and J.D. Skinner (2009). 'An allometric analysis of the giraffe cardiovascular system', *Comparative Biochemistry and Physiology Part A: Molecular & Integrative Physiology*, Vol. 154, Issue 4, pp. 523–29. https://doi.org/10.1016/j.cbpa.2009.08.013

6 Damkjær, M., T. Wang, E. Brøndum, K.H. Østergaard, U. Baandrup, A. Hørlyck, J.M. Hasenkam, M. Smerup, J. Funder, N. Marcussen, and C.C. Danielsen (2015). 'The giraffe kidney tolerates high arterial blood pressure by high renal interstitial pressure and low glomerular filtration rate', *Acta Physiologica*, Vol. 214, Issue 4, pp.497–510. https://doi.org/10.1111/apha.12531

7 Paton, J.F.R., C.J. Dickinson, and G. Mitchell (2009). 'Harvey Cushing and the regulation of blood pressure in giraffe, rat and man: introducing "Cushing's mechanism"', *Experimental Physiology*, Vol. 94, No. 1, pp. 11–17. https://doi.org/10.1113/expphysiol.2008.043455

8 Solounias, N. (1999). 'The remarkable anatomy of the giraffe's neck', *Journal of Zoology*, Vol. 247, Issue 2, pp. 257–68. https://doi.org/10.1111/j.1469-7998.1999.tb00989.x

9 Endo, H., D. Yamagiwa, M. Fujisawa, J. Kimura, M. Kurohmaru, and Y. Hayashi (1997). 'Modified neck muscular system of the giraffe (Giraffa camelopardalis)'. *Annals of Anatomy-Anatomischer Anzeiger*, Vol. 179, Issue 5, pp. 481–85. https://doi.org/10.1016/S0940-9602(97)80055-6

29. Airborne Sleep

1 Rattenborg, N.C., B. Voirin, S.M. Cruz, R. Tisdale, G. Dell'Omo, H.P. Lipp, M. Wikelski, and A.L. Vyssotski (2016). 'Evidence that birds sleep in mid-flight', *Nature Communications*, Vol. 7, Article No. 12468. https://doi.org/10.1038/ncomms12468

2 Rattenborg, N.C. (2006). 'Do birds sleep in flight?' *Naturwissenschaften*, Vol. 93, No. 9, pp. 413–25. https://doi.org/10.1007/s00114-006-0120-3

3 Weimerskirch, H., C. Bishop, T. Jeanniard-du-Dot, A. Prudor, and G. Sachs (2016). 'Frigate birds track atmospheric conditions over months-long transoceanic flights', *Science*, Vol. 353, Issue 6294, pp. 74–78. https://doi.org/10.1126/science.aaf4374

4 Rattenborg, N.C. (2017). 'Sleeping on the wing', *Interface Focus*, Vol. 7, Issue 1, 20160082. https://doi.org/10.1098/rsfs.2016.0082

30. Learning from Monkeys?

1 Haslam, M., L.V. Luncz, R.A. Staff, F. Bradshaw, E.B. Ottoni, and T. Falótico (2016). 'Pre-Columbian monkey tools', *Current Biology*, Vol. 26, Issue 13, R521–R522. https://doi.org/10.1016/j.cub.2016.05.046

2 Visalberghi, E., A. Albani, M. Ventricelli, P. Izar, G. Schino, and D. Fragazsy (2016). 'Factors affecting cashew processing by wild bearded capuchin monkeys (Sapajus libidinosus, Kerr 1792)', *American Journal of Primatology*, Vol. 78, Issue 8, pp. 799–815. https://doi.org/10.1002/ajp.22545

3 Spagnoletti, N., E. Visalberghi, E. Ottoni, P. Izar, and D. Fragaszy (2011). 'Stone tool use by adult wild bearded capuchin monkeys (Cebus libidinosus): frequency, efficiency and tool selectivity', *Journal of Human Evolution*, Vol. 61, Issue 1, pp. 97–107. https://doi.org/10.1016/j.jhevol.2011.02.010

4 Fragaszy, D.M., Q. Liu, B.W. Wright, A. Allen, C.W. Brown, and E. Visalberghi (2013). 'Wild bearded capuchin monkeys (Sapajus libidinosus) strategically place nuts in a stable position during nut-cracking', *PLoS One*, Vol. 8, Issue 2, e56182. https://doi.org/10.1371/journal.pone.0056182

5 Falótico, T., L.V. Luncz, M.S. Svensson, and M. Haslam (2016). 'Cashew nut positioning during stone tool use by wild bearded capuchin monkeys (Sapajus libidinosus)', *Folia Primatologica*, Vol. 87, Issue 6, pp. 392–97. https://doi.org/10.1159/000459621

6 Mercader, J., H. Barton, J. Gillespie, J. Harris, S. Kuhn, R. Tyler, and C. Boesch (2007). '4,300-year-old chimpanzee sites and the origins of percussive stone technology', *Proceedings of the National Academy of Sciences*, Vol. 104, Issue 9, pp. 3043–48. https://doi.org/10.1073/pnas.0607909104

7 Haslam, M., L. Luncz, A. Pascual-Garrido, T. Falótico, S. Malaivijitnond, and M. Gumert (2016). 'Archaeological excavation of wild macaque stone tools', *Journal of Human Evolution*, No. 96, pp. 134–38. https://doi.org/10.1016/j.jhevol.2016.05.002

8 Falótico, T., and E.B. Ottoni (2016). 'The manifold use of pounding stone tools by wild capuchin monkeys of Serra da Capivara National Park, Brazil', *Behaviour*, Vol. 153, Issue 4, pp. 421–42. https://doi.org/10.1163/1568539X-00003357

9 Moura, A. C. D. A. (2007). 'Stone banging by wild capuchin monkeys: an unusual auditory display', *Folia Primatologica*, Vol. 78, Issue 1, pp. 36–45. https://doi.org/10.1159/000095684

10 Falótico, T., and E.B. Ottoni (2013). 'Stone throwing as a sexual display in wild female bearded capuchin monkeys, *Sapajus libidinosus*', *PLoS One*, Vol. 8, Issue 11, e79535. https://doi.org/10.1371/journal.pone.0079535

11 Haslam, M., and T. Falótico (2015). 'Nasal probe and toothpick tool use by a wild female bearded capuchin (*Sapajus libidinosus*)', *Primates*, Vol. 56, Issue 3, pp. 211–14. https://doi.org/10.1007/s10329-015-0470-6

31. Honey-Collecting with Birds

1 Spottiswoode, C.N., K.S. Begg, and C.M. Begg (2016). 'Reciprocal signaling in honeyguide-human mutualism', *Science*, Vol. 353, Issue 6297, pp. 387–89. https://doi.org/10.1126/science.aaf4885

2 Isack, H.A., and H.U. Reyer (1989). 'Honeyguides and honey gatherers: interspecific communication in a symbiotic relationship', *Science*, Vol. 243, Issue 4896, pp. 1343–46. https://doi.org/10.1126/science.243.4896.1343

3 Dean, W.R.J., W.R. Siegfried, and I.A.W. MacDonald (1990). 'The fallacy, fact, and fate of guiding behavior in the greater honeyguide', *Conservation Biology*, Vol. 4, Issue 1, pp. 99–101. https://doi.org/10.1111/j.1523-1739.1990.tb00272.x

4 Wood, B.M., H. Pontzer, D.A. Raichlen, and F.W. Marlowe (2014). 'Mutualism and manipulation in Hadza–honeyguide interactions', *Evolution and Human Behavior*, Vol. 35, Issue 6, pp. 540–46. https://doi.org/10.1016/j.evolhumbehav.2014.07.007

5 Fincham, J.E., R. Peek, and M. Markus (2017). 'The Greater Honeyguide: Reciprocal signalling and innate recognition of a Honey Badger', *Biodiversity Observations*, Vol. 8, Issue12, 1–6.

32. Are Humpback Whales Altruistic?

1 Pitman, R.L., V.B. Deecke, C.M. Gabriele, M. Srinivasan, N. Black, J. Denkinger, J.W. Durban, E.A. Mathews, D.R. Matkin, J.L. Neilson, and A. Schulman-Janiger, (2017). 'Humpback whales interfering when mammal-eating killer whales attack other species: Mobbing behavior and interspecific altruism?', *Marine Mammal Science*, Vol. 33, Issue 1, pp. 7–58. https://doi.org/10.1111/mms.12343

2 Mehta, A.V., J.M. Allen, R. Constantine, C. Garrigue, B. Jann, C. Jenner, M.K. Marx, C.O. Matkin, D.K. Mattila, G. Minton, and

S.A. Mizroch (2007). 'Baleen whales are not important as prey for killer whales Orcinus orca in high-latitude regions', *Marine Ecology Progress Series*, No. 348, pp. 297–307. https://doi.org/10.3354/meps07015

3 Pitman, R.L., J.A. Totterdell, H. Fearnbach, L.T. Ballance, J.W. Durban, and H. Kemps (2015). 'Whale killers: prevalence and ecological implications of killer whale predation on humpback whale calves off Western Australia', *Marine Mammal Science*, Vol. 31, Issue 2, pp. 629–57. https://doi.org/10.1111/mms.12182

33. The Playful Keas of New Zealand

1 Schwing, R., X.J. Nelson, A. Wein, and S. Parsons (2017). 'Positive emotional contagion in a New Zealand parrot', *Current Biology*, Vol. 27, Issue 6, R213–R214. https://doi.org/10.1016/j.cub.2017.02.020

2 Diamond, J., D. Eason, C. Reid, and A.B. Bond (2006). 'Social play in kakapo (*Strigops habroptilus*) with comparisons to kea (*Nestor notabilis*) and kaka (*Nestor meridionalis*). *Behaviour*, Vol. 143, Issue 11, pp. 1397–1423. http://doi.org/10.1163/156853906778987551

34. Peeing on Eggs

1 Delia, J., L. Bravo-Valencia, and K.M. Warkentin (2017). 'Patterns of parental care in Neotropical glass frogs: fieldwork alters hypotheses of sex-role evolution', *Journal of Evolutionary Biology*, Vol. 30, Issue 5, pp. 898–914. https://doi.org/10.1111/jeb.13059

2 Dautel, N., A.L.S. Maldonado, R. Abuza, H. Imba, K. Griffin, and J.M. Guayasamin (2011). 'Advertisement and combat calls of the glass frog Centrolene lynchi (*Anura: Centrolenidae*), with notes on combat and reproductive behaviors', *Phyllomedusa: Journal of Herpetology*, Vol. 10, No. 1, pp. 31–43. https://doi.org/10.11606/issn.2316-9079.v10i1p31-43

3 Delia, J.R., A. Ramírez-Bautista, and K. Summers (2014). 'Glass frog embryos hatch early after parental desertion', *Proceedings of the Royal Society B: Biological Sciences*, Vol. 281, Issue 1785, 20133237. https://doi.org/10.1098/rspb.2013.3237

35. No Ants Left Behind

1 Frank, E.T., T. Schmitt, T. Hovestadt, O. Mitesser, J. Stiegler, and K.E. Linsenmair (2017). 'Saving the injured: rescue behavior in the termite-hunting ant *Megaponera analis*', *Science Advances*, Vol. 3, No. 4, e1602187. https://doi.org/10.1126/sciadv.1602187

2 Frank, E.T., M. Wehrhahn, and K.E. Linsenmair (2018). 'Wound treatment and selective help in a termite-hunting ant', *Proceedings of the Royal Society B: Biological Sciences*, Vol. 285, Issue 1872, 20172457. https://doi.org/10.1098/rspb.2017.2457

36. Feigning Death to Avoid Sex?

1 Khelifa, R. (2017). 'Faking death to avoid male coercion: extreme sexual conflict resolution in a dragonfly', *Ecology*, Vol. 98, Issue 6, pp. 1724–26. https://doi.org/10.1002/ecy.1781

2 Rogers, S.M., and S.J. Simpson (2014). 'Thanatosis', *Current Biology*, Vol. 24, Issue 21, R1031–R1033. https://doi.org/10.1016/j.cub.2014.08.051

3 Bilde, T., C. Tuni, R. Elsayed, S. Pekár, and S. Toft (2005). 'Death feigning in the face of sexual cannibalism', *Biology Letters*, Vol. 2, Issue 1, pp. 23–25. https://doi.org/10.1098/rsbl.2005.0392

4 Rüppell, G., and D. Hilfert-Rüppell (2014). 'Slow-motion analysis of female refusal behaviour in dragonflies', *International Journal of Odonatology*, Vol. 17, Issue 4, pp. 199–215. https://doi.org/10.1080/13887890.2014.972893

5 Janssenswillen, S., and F. Bossuyt (2016). 'Male courtship pheromones induce cloacal gaping in female newts (Salamandridae)' *PloS One*, Vol. 11, No. 1, e0144985. https://doi.org/10.1371/journal.pone.0144985

6 Córdoba-Aguilar, A. (ed.). (2008). *Dragonflies and Damselflies: Model organisms for ecological and evolutionary research,* Oxford: OUP.

7 Eberhard, W.G. (2002). 'The function of female resistance behavior: intromission by male coercion vs. female cooperation in sepsid flies

(*Diptera: Sepsidae*)', *Revista de Biologia Tropical*, Vol. 50, Issue 2, pp. 485–505.

37. Scolding Egyptian Fruit Bats

1 Prat, Y., M. Taub, and Y. Yovel (2016). 'Everyday bat vocalizations contain information about emitter, addressee, context, and behavior', *Scientific Reports*, 6, Article No. 39419. https://doi.org/10.1038/srep39419

2 Prat, Y., L. Azoulay, R. Dor, and Y. Yovel (2017). 'Crowd vocal learning induces vocal dialects in bats: playback of conspecifics shapes fundamental frequency usage by pups', *PLoS biology*, Vol. 15, Issue 10, e2002556. https://doi.org/10.1371/journal.pbio.2002556

38. Music in the Wild

1 Heinsohn, R., C.N. Zdenek, R.B. Cunningham, J.A. Endler, and N.E. Langmore (2017). 'Tool-assisted rhythmic drumming in palm cockatoos shares key elements of human instrumental music', *Science Advances*, Vol. 3, No. 6, e1602399. https://doi.org/10.1126/sciadv.1602399

2 Wood, G.A. (1984). 'Tool Use by the Palm Cockatoo *Probosciger aterrimus* During Display', *Corella,* Vol. 8, Issue 4, pp. 94–95. https://absa.asn.au/post_documents/tool-use-by-the-palm-cockatoo-probosciger-atriums-during-display/

3 Wood, G.A. (1988). 'Further Field Observations of the Palm Cockatoo (*Probosciger aterrimus*) in the Cape York Peninsula, Queensland', *Corella* Vol. 12, No. 2, pp. 48–52. http://aainsects.com.au/Pdf-papers-aainsects/Bird/bird-Cacat-3.pdf

4 Murphy, S., S. Legge, and R. Heinsohn (2003). 'The breeding biology of palm cockatoos (*Probosciger aterrimus*): a case of a slow life history', *Journal of Zoology*, Vol. 261, Issue 4, pp. 327–39. https://doi.org/10.1017/S0952836903004175

5 Patel, A.D. (2014). 'The evolutionary biology of musical rhythm: was Darwin wrong?' *PLoS Biology*, Vol. 12, Issue 3, e1001821. https://doi.org/10.1371/journal.pbio.1001821

6 Arcadi, A.C., D. Robert, and C. Boesch (1998). 'Buttress drumming by wild chimpanzees: temporal patterning, phrase integration into loud calls, and preliminary evidence for individual distinctiveness', *Primates*, Vol. 39, Issue 4, pp. 505–18. https://doi.org/10.1007/BF02557572

7 Hattori, Y., M. Tomonaga, and T. Matsuzawa (2013). 'Spontaneous synchronized tapping to an auditory rhythm in a chimpanzee', *Scientific Reports*, No. 3, Article No. 1566. https://doi.org/10.1038/srep01566

8 Large, E.W., and P.M. Gray (2015). 'Spontaneous tempo and rhythmic entrainment in a bonobo (Pan paniscus)', *Journal of Comparative Psychology*, Vol. 129, No. 4, pp. 317–28. http://doi.org/10.1037/com0000011

9 Dufour, V., N. Poulin, C. Curé, and E.H. Sterck (2015). 'Chimpanzee drumming: a spontaneous performance with characteristics of human musical drumming', *Scientific Reports*, 5, Article No. 11320. https://doi.org/10.1038/srep11320

39. Land Snails' Secret Weapons

1 Rae, R. (2017). 'The gastropod shell has been co-opted to kill parasitic nematodes', *Scientific Reports* 7, Article No. 4745. https://doi.org/10.1038/s41598-017-04695-5

2 Williams, A., and R. Rae (2016). 'Cepaea nemoralis (Linnaeus, 1758) uses its shell as a defence mechanism to trap and kill parasitic nematodes', *Journal of Molluscan Studies*, Vol. 82, Issue 2, pp. 349–50. https://doi.org/10.1093/mollus/eyv064

3 Williams, A. J., and R. Rae (2015). 'Susceptibility of the Giant African snail (Achatina fulica) exposed to the gastropod parasitic nematode Phasmarhabditis hermaphrodita', *Journal of Invertebrate Pathology*, Vol. 127, pp. 122–26. https://doi.org/10.1016/j.jip.2015.03.012

4 Rae, R. (2017). 'How I showed that snails use their shells to trap and kill parasites', *The Conversation*, 6 July 6 2017. https://theconversation.com/how-i-showed-that-snails-use-their-shells-to-trap-and-kill-parasites-80564
5 De Baets, K., C. Klug, and D. Korn (2011). 'Devonian pearls and ammonoid-endoparasite co-evolution', *Acta Palaeontologica Polonica*, Vol. 56, No. 1, pp. 159–80. https://doi.org/10.4202/app.2010.0044

40. Sex Drive in Fish

1 Sommer-Trembo, C., M. Plath, J. Gismann, C. Helfrich, and D. Bierbach (2017). 'Context-dependent female mate choice maintains variation in male sexual activity', *Royal Society Open Science*, Vol. 4, Issue 7, 170303. https://doi.org/10.1098/rsos.170303
2 Dadda, M. (2015). 'Female social response to male sexual harassment in poeciliid fish: a comparison of six species' *Frontiers in Psychology*, 6, Article No. 1453. https://doi.org/10.3389/fpsyg.2015.01453
3 Pilastro, A., E. Giacomello, and A. Bisazza (1997). 'Sexual selection for small size in male mosquitofish (Gambusia holbrooki)', *Proceedings of the Royal Society of London. Series B: Biological Sciences*, Vol. 264, Issue 1385, pp. 1125–29. https://doi.org/10.1098/rspb.1997.0155
4 Greven, H. (2005). 'Structural and behavioral traits associated with sperm transfer in Poeciliinae', in Uribe, M.C. and H.J. Grier (eds) (2010). *Viviparous Fishes*, Homestead, Florida: New Life Publications, pp 145–63. https://www.researchgate.net/profile/Hartmut_Greven/publication/259266955_Structural_and_behavioural_traits_associated_with_sperm_transfer_in_Poeciliinae/links/0046352aa30cec830f000000.pdf
5 Dadda, M. (2015). 'Female social response to male sexual harassment in poeciliid fish: a comparison of six species', *Frontiers in Psychology*, 6, Article No. 1453. https://doi.org/10.3389/fpsyg.2015.01453
6 Culumber, Z.W., and M. Tobler (2017). 'Sex-specific evolution during the diversification of live-bearing fishes', *Nature, Ecology*

& Evolution, Vol. 1, No. 8, pp. 1185–91. https://doi.org/10.1038/s41559-017-0233-4

41. Voting in African Wild Dogs

1 Walker, R.H., A.J. King, J.W. McNutt, and N.R. Jordan (2017). 'Sneeze to leave: African wild dogs (Lycaon pictus) use variable quorum thresholds facilitated by sneezes in collective decisions', *Proceedings of the Royal Society B: Biological Sciences*, Vol. 284, No. 1862, 2017.0347. https://doi.org/10.1098/rspb.2017.0347

2 Estes, R.D., and J. Goddard (1967). 'Prey selection and hunting behavior of the African wild dog', *The Journal of Wildlife Management*, Vol. 31, No. 1, pp. 52–70. https://doi.org/10.2307/3798360

3 Cohen, J. A., and M.W. Fox (1976). 'Vocalizations in wild canids and possible effects of domestication', *Behavioural Processes* Vol. 1, Issue 1, pp. 77–92. https://doi.org/10.1016/0376-6357(76)90008-5

4 Lehner, P.N. (1978). 'Coyote vocalizations: a lexicon and comparisons with other canids', *Animal Behaviour*, No. 26, Part 3, pp. 712–22. https://doi.org/10.1016/0003-3472(78)90138-0

5 Smith, B. (ed.) (2015). *The Dingo Debate: Origins, Behaviour and Conservation*, Clayton, Victoria, Australia: CSIRO Publishing.

6 Bousquet, C.A., D.J. Sumpter, and M.B. Manser (2010). 'Moving calls: a vocal mechanism underlying quorum decisions in cohesive groups', *Proceedings of the Royal Society B: Biological Sciences*, Vol. 278, Issue 1711, pp. 1482–88. https://doi.org/10.1098/rspb.2010.1739

42. Locking Poisons Out

1 http://www.toxinology.com/terrestrial_vertebrates/ns-treat-57.html –Accessed on 27 April 2019.

2 Tarvin, R.D., J.C. Santos, L.A. O'Connell, H.H. Zakon, and D.C. Cannatella (2016). 'Convergent substitutions in a sodium channel suggest multiple origins of toxin resistance in poison frogs', *Molecular Biology and Evolution*, Vol. 33, Issue 4, pp. 1068–80. https://doi.org/10.1093/molbev/msv350

3 Wang, S., and G.K. Wang (2017). 'Single rat muscle Na+ channel mutation confers batrachotoxin autoresistance found in poison-dart frog Phyllobates terribilis', *Proceedings of the National Academy of Sciences*, Vol. 114, No. 39, pp. 10491–96. https://doi.org/10.1073/pnas.1707873114

4 Tarvin, R.D., C.M. Borghese, W. Sachs, J.C. Santos, Y. Lu, L.A. O'Connell, D.C. Cannatella, R.A. Harris, and H.H. Zakon (2017). 'Interacting amino acid replacements allow poison frogs to evolve epibatidine resistance', *Science*, Vol. 357, Issue 6357, pp. 1261–66. https://doi.org/10.1126/science.aan5061

5 Traynor, J.R. (1998). 'Epibatidine and pain', *British Journal of Anaesthesia*, Vol. 81, Issue 1, pp. 69–76. https://doi.org/10.1093/bja/81.1.69

6 Márquez, R., V. Ramírez-Castañeda, and A. Amézquita (2019). 'Does batrachotoxin autoresistance coevolve with toxicity in Phyllobates poison-dart frogs?' *Evolution*, Vol. 73, Issue 2, pp. 390–400. https://doi.org/10.1111/evo.13672

43. Laziness Has Its Uses

1 Charbonneau, D., T. Sasaki, and A. Dornhaus (2017). 'Who needs "lazy" workers? Inactive workers act as a "reserve" labor force replacing active workers, but inactive workers are not replaced when they are removed', *Plos One*, Vol. 12, Issue 9, e0184074. https://doi.org/10.1371/journal.pone.0184074

2 Charbonneau, D., N. Hillis, and A., Dornhaus (2015). 'Lazy' in nature: ant colony time budgets show high "inactivity" in the field as well as in the lab', *Insectes Sociaux*, Vol. 62, No. 1, pp. 31–35. https://doi.org 10.1007/s00040-014-0370-6

3 Charbonneau, D., C. Poff, H. Nguyen, M.C. Shin, K. Kierstead and A. Dornhaus (2017). 'Who are the "lazy" ants? The function of inactivity in social insects and a possible role of constraint: inactive ants are corpulent and may be young and/or selfish', *Integrative and Comparative Biology*, Vol. 57, Issue 3, pp. 649–67. https://doi.org/10.1093/icb/icx029

4 Charbonneau, D., C. Poff, H. Nguyen, M.C. Shin, K. Kierstead and A. Dornhaus (2017). 'Who are the "lazy" ants? The function of inactivity in social insects and a possible role of constraint: inactive ants are corpulent and may be young and/or selfish', *Integrative and Comparative Biology*, Vol. 57, Issue 3, pp. 649–67. https://doi.org/10.1093/icb/icx029

5 Gardner, K.E., R.L. Foster, and S. O'Donnell (2007). 'Experimental analysis of worker division of labor in bumblebee nest thermoregulation (Bombus huntii, Hymenoptera: Apidae)', *Behavioral Ecology and Sociobiology*, Vol. 61, No. 5, pp. 783–92. https://doi.org/10.1007/s00265-006-0309-7

6 Ishii, Y., and E. Hasgeawa (2013). 'The mechanism underlying the regulation of work-related behaviors in the monomorphic ant, Myrmica kotokui', *Journal of Ethology*, Vol. 31, No. 1, pp. 61–69. https://doi.org/10.1007/s10164-012-0349-6

7 Yamamoto, T., and E. Hasegawa (2017). 'Response threshold variance as a basis of collective rationality. *Royal Society Open Science*, Vol. 4, Issue 4, 170097. https://doi.org/10.1098/rsos.170097

8 Hasegawa, E., Y. Ishii, K. Tada, K. Kobayashi, and J, Yoshimura (2016). 'Lazy workers are necessary for long-term sustainability in insect societies', *Scientific Reports*, No. 6, Article No. 20846. https://doi.org/10.1038/srep20846

44. Shrinking Shrews

1 Lázaro, J., D.K. Dechmann, S. LaPoint, M. Wikelski, and M. Hertel (2017). 'Profound reversible seasonal changes of individual skull size in a mammal', *Current Biology*, Vol. 27, Issue 20, R1106–R1107. https://doi.org/10.1016/j.cub.2017.08.055

2 Taylor, J.R., L. Rychlik, and S. Churchfield (2012). 'Winter reduction in body mass in a very small, non-hibernating mammal: consequences for heat loss and metabolic rates', *Physiological and Biochemical Zoology*, Vol. 86, No. 1, pp. 9–18. https://doi.org/10.1086/668484

3 Heldmaier, G. (1989). 'Seasonal acclimatization of energy requirements in mammals: functional significance of body

weight control, hypothermia, torpor and hibernation', *Energy Transformations in Cells and Organisms*, pp.130–39. https://www.researchgate.net › publication › 235123255

4 Ochocińska, D., and J.R. Taylor (2003). 'Bergmann's rule in shrews: geographical variation of body size in Palearctic Sorex species', *Biological Journal of the Linnean Society*, Vol. 78, Issue 3, pp. 365–81. https://doi.org/10.1046/j.1095-8312.2003.00150.x

5 Pucek, Z. (1963). 'Seasonal changes in the braincase of some representatives of the genus Sorex from the Palearctic', *Journal of Mammalogy*, Vol. 44, Issue 4, pp. 523–36. https://doi.org/10.2307/1377135

6 Grigoriev, A.I., V.S. Oganov, A.V. Bakulin, V.V. Poliakov, L.I. Voronin, V.V. Morgun, V.S. Shnaĭder, L.V. Murashko, V.E. Novikov, A. LeBlank, and L. Shakleford (1998). 'Clinical and physiological evaluation of bone changes among astronauts after long-term space flights', *Aviakosmicheskaia i ekologicheskaia meditsina*= Aerospace and environmental medicine, Vol. 32, No. 1, pp. 21–25.

7 Smith, S.M., M.A. Heer, L.C. Shackelford, J.D. Sibonga, L. Ploutz-Snyder, and S.R. Zwart (2012). 'Benefits for bone from resistance exercise and nutrition in long-duration spaceflight: evidence from biochemistry and densitometry', *Journal of Bone and Mineral Research*, Vol. 27, No. 9, pp. 1896–1906. https://doi.org/10.1002/jbmr.1647

8 Dechmann, D.K., S. LaPoint, C. Dullin, M. Hertel, J.R. Taylor, K. Zub, and M. Wikelski (2017). 'Profound seasonal shrinking and regrowth of the ossified braincase in phylogenetically distant mammals with similar life histories', *Scientific Reports*, No. 7, 42443. https://doi.org/10.1038/srep42443

9 Lázaro, J., M. Hertel, C.C. Sherwood, M. Muturi, and D.K. Dechmann (2018). 'Profound seasonal changes in brain size and architecture in the common shrew', *Brain Structure and Function*, Vol. 223, No. 6, pp. 2823–40. https://doi.org/10.1007/s00429-018-1666-5

10 Lázaro, J., M. Hertel, S. LaPoint, M. Wikelski, M. Stiehler, and D.K. Dechmann (2018). 'Cognitive skills of common shrews (Sorex

araneus) vary with seasonal changes in skull size and brain mass', *Journal of Experimental Biology*, Vol. 221, No. 2, jeb166595. https://doi.org/10.1242/jeb.166595

45. Wading Birds in the Desert

1 Pedler, R.D., R.F. Ribot, and A.T. Bennett (2018). 'Long-distance flights and high-risk breeding by nomadic waterbirds on desert salt lakes', *Conservation Biology*, Vol. 32, Issue 1, pp. 216–28. https://doi.org/10.1111/cobi.13007

2 McGilp, J.N., and A.M. Morgan (1931). 'The Nesting of the Banded Stilt', *The South Australian Ornithologist*, Vol. XI, Part 2, pp. 37–53. https://birdssa.asn.au/images/saopdfs/Volume11/1931V11P037.pdf

3 Williams, W.D., P. De Deckker, and R.J. Shiel (1998). 'The limnology of Lake Torrens, an episodic salt lake of central Australia, with particular reference to unique events in 1989', *Hydrobiologia*, Vol. 384, Issue 1-3, pp. 101–10. https://doi.org/10.1023/A:1003207613473

4 Pedler, R.D., R.F.H. Ribot, and A.T.D. Bennett (2014). 'Extreme nomadism in desert waterbirds: flights of the banded stilt', *Biology Letters*, Vol. 10, Issue 10, 20140547. https://doi.org/10.1098/rsbl.2014.0547

5 Poiani, A. (2006). 'Effects of floods on distribution and reproduction of aquatic birds', *Advances in Ecological Research*, Vol. 39, pp. 63–83. https://doi.org/10.1016/S0065-2504(06)39004-6

6 Kingsford, R.T., D.A. Roshier, and J.L. Porter (2010). 'Australian waterbirds – time and space travellers in dynamic desert landscapes', *Marine and Freshwater Research*, Vol. 61, No. 8, pp. 875–84. https://doi.org/10.1071/MF09088

7 Adamou, A.E., M. Kouidri, Y. Chabi, J. Skwarska, and J. Bańbura (2009). 'Egg size variation and breeding characteristics of the Black-winged Stilt Himantopus himantopus in a Saharan oasis', *Acta Ornithologica*, Vol. 44, Issue 1, pp. 1–7. https://doi.org/10.3161/000164509X464821

46. Bird Nannies

1 Kingma, S.A. (2017). 'Direct benefits explain interspecific variation in helping behaviour among cooperatively breeding birds', *Nature Communications*, Vol. 8, Article No 1094. https://doi.org/10.1038/s41467-017-01299-5

2 Fessl, B., S. Kleindorfer, and H. Hoi (1996). 'Extra male parental behaviour: evidence for an alternative mating strategy in the Moustached Warbler Acrocephalus melanopogon', *Journal of Avian Biology*, Vol. 27, No. 1, pp. 88-91. https://doi.org/10.2307/3676965

3 Dunn, P.O., A. Cockburn, and R.A. Mulder (1995). 'Fairy-wren helpers often care for young to which they are unrelated', *Proceedings of the Royal Society of London. Series B: Biological Sciences*, Vol. 259, Issue 1356, pp. 339–43. https://doi.org/10.1098/rspb.1995.0050

4 Painter, J.N., R.H. Crozier, A. Poiani, R.J. Robertson, and M.F. Clarke (2000). 'Complex social organization reflects genetic structure and relatedness in the cooperatively breeding bell miner, Manorina melanophrys', *Molecular Ecology*, Vol. 9, Issue 9, pp. 1339–47. https://doi.org/10.1046/j.1365-294x.2000.01012.x

5 Russell, A.F., N.E. Langmore, A. Cockburn, L.B. Astheimer, and R.M. Kilner (2007). 'Reduced egg investment can conceal helper effects in cooperatively breeding birds', *Science*, Vol. 317, Issue 5840, pp. 941–44. https://doi.org/10.1126/science.1146037

6 Conrad, K.F., M.F. Clarke, R.J. Robertson, and P.T. Boag (1998). 'Paternity and the relatedness of helpers in the cooperatively breeding bell miner', *The Condor*, Vol. 100, Issue 2, pp. 343–49. https://doi.org/10.2307/1370275

7 Riehl, C. (2013). 'Evolutionary routes to non-kin cooperative breeding in birds', *Proceedings of the Royal Society B: Biological Sciences*, Vol. 280, Issue 1772, 20132245. https://doi.org/10.1098/rspb.2013.2245

8 Reyer, H.U. (1986). 'Breeder-helper-interactions in the pied kingfisher reflect the costs and benefits of cooperative breeding', *Behaviour*, Vol. 96, Issue 3–4, pp. 277–302. https://doi.org/10.1163/156853986X00522

9 Cockburn, A. (1998). 'Evolution of helping behavior in cooperatively breeding birds', *Annual Review of Ecology and Systematics*, Vol. 29, No. 1, pp. 141–77. https://doi.org/10.1146/annurev.ecolsys.29.1.141

10 Sherman, P.T. (1995). 'Social organization of cooperatively polyandrous white-winged trumpeters (Psophia leucoptera)', *The Auk*, Vol. 112, No. 2, pp. 296–309.

47. Dads vs. Cannibals

1 Seshadri, K.S., and D.P. Bickford (2018). 'Faithful fathers and crooked cannibals: the adaptive significance of parental care in the bush frog Raorchestes chalazodes, Western Ghats, India', *Behavioral Ecology and Sociobiology*, Vol. 72, Article No. 4. https://doi.org/10.1007/s00265-017-2420-3

2 Seshadri, K.S., K.V. Gururaja, and N.A. Aravind (2012). 'A new species of Raorchestes (Amphibia: Anura: Rhacophoridae) from mid-elevation evergreen forests of the southern Western Ghats, India', *Zootaxa*, No. 3410, No. 1, pp. 19–34.

3 Gururaja, K.V., K.P. Dinesh, M.J. Palot, C. Radhakrishnan, and T.V. Ramachandra (2007). 'A new species of Philautus Gistel (Amphibia: Anura: Rhacophoridae) from southern Western Ghats, India', *Zootaxa*, Vol. 1621, No. 1, pp. 1–16.

4 Seshadri, K.S., K.V. Gururaja, and D.P. Bickford (2015). 'Breeding in bamboo: a novel anuran reproductive strategy discovered in Rhacophorid frogs of the Western Ghats, India', *Biological Journal of the Linnean Society*, Vol. 114, No. 1, pp. 1–11. https://doi.org/10.1111/bij.12388

48. High Altitude Survival

1 Lague, S.L., B. Chua, L. Alza, G.R. Scott, P.B. Frappell, Y. Zhong, A.P. Farrell, K.G. McCracken, Y. Wang, and W.K. Milsom (2017). 'Divergent respiratory and cardiovascular responses to hypoxia in bar-headed geese and Andean birds', *Journal of Experimental Biology*, Vol. 220, pp. 4186–94. https://doi.org/10.1242/jeb.168799

2 Hawkes, L.A., S. Balachandran, N. Batbayar, P.J. Butler, B. Chua, D.C. Douglas, P.B. Frappell, Y. Hou, W.K. Milsom, S.H. Newman, and D.J. Prosser (2013). 'The paradox of extreme high-altitude migration in bar-headed geese Anser indicus', *Proceedings of the Royal Society B: Biological Sciences*, Vol. 280, Issue 1750, 20122114. https://doi.org/10.1098/rspb.2012.2114

3 Bishop, C.M., R.J. Spivey, L.A. Hawkes, N. Batbayar, B. Chua, P.B. Frappell, W.K. Milsom, T. Natsagdorj, S.H. Newman, G.R. Scott, and J.Y. Takekawa (2015). 'The roller-coaster flight strategy of bar-headed geese conserves energy during Himalayan migrations', *Science*, Vol, 347, Issue 6219, pp. 250–54. https://doi.org/10.1126/science.1258732

4 Hawkes, L.A., S. Balachandran, N. Batbayar, P.J. Butler, P.B. Frappell, W.K. Milsom, N. Tseveenmyadag, S.H. Newman, G.R. Scott, P. Sathiyaselvam, and J.Y.Takekawa (2011). 'The trans-Himalayan flights of bar-headed geese (Anser indicus)', *Proceedings of the National Academy of Sciences*, Vol. 108, No. 23, pp. 9516–19. https://doi.org/10.1073/pnas.1017295108

5 Maina, J.N., K.G. McCracken, B. Chua, J.M. York, and W.K. Milsom (2017). 'Morphological and morphometric specializations of the lung of the Andean goose, Chloephaga melanoptera: A lifelong high-altitude resident', *Plos One*, Vol. 12, Issue 3, e0174395. https://doi.org/10.1371/journal.pone.0174395

6 Faraci, F.M., and M.R. Fedde (1986). 'Regional circulatory responses to hypocapnia and hypercapnia in bar-headed geese', *American Journal of Physiology-Regulatory, Integrative and Comparative Physiology*, Vol. 250, Issue 3, pp. R499–R504. https://doi.org/10.1152/ajpregu.1986.250.3.R499

7 Beall, C.M. (2007). 'Two routes to functional adaptation: Tibetan and Andean high-altitude natives', *Proceedings of the National Academy of Sciences*, Vol. 104, Suppl. 1, pp. 8655–60. https://doi.org/10.1073/pnas.0701985104

8 Beall, C. M., D. Laskowski, and S.C. Erzurum (2012). 'Nitric oxide in adaptation to altitude', *Free Radical Biology and Medicine*,

Vol. 52, Issue 7, pp. 1123–34. https://doi.org/10.1016/j.freeradbiomed.2011.12.028

9 Dempsey, J.A., and B.J. Morgan (2015). 'Humans in hypoxia: a conspiracy of maladaptation?!', *Physiology*, Vol. 30, Issue 4, pp. 304–16. https://doi.org/10.1152/physiol.00007.2015

49. Firehawks

1 Bonta, M., R. Gosford, D. Eussen, N. Ferguson, E. Loveless, and M. Witwer (2017). 'Intentional fire-spreading by "Firehawk" raptors in Northern Australia', *Journal of Ethnobiology*, Vol. 37, No. 4, pp. 700–719. https://doi.org/10.2993/0278-0771-37.4.700

50. Devoted Snake Moms

1 Alexander, G.J. (2018). 'Reproductive biology and maternal care of neonates in southern African python (Python natalensis)'. *Journal of Zoology*, Vol. 305, Issue 3, pp. 141–48. https://doi.org/10.1111/jzo.12554

2 Ramesh, C., and S. Bhupathy (2010). 'Breeding biology of Python molurus molurus in Keoladeo national park, Bharatpur, India', *The Herpetological Journal*, Vol. 20, No. 3, pp. 157–63.

3 Greene, H.W., P.G. May, D.L. Hardy Sr, J.M. Sciturro, and T.M. Farrell (2002). 'Parental behavior by vipers', *Biology of the Vipers*, pp. 179–205. http://eaglemountainpublishing.s3.amazonaws.com/PDF/Biology%20of%20the%20Vipers/CH%2012_greene_.pdf

4 Alexander, G. Personal communication.

Index

Acknowledgements

THIS collection wouldn't have come into being without Mukunth Vasudevan, the science editor at *The Wire*. I subscribe to alerts from scientific journals that I would scan every evening after a long day's work. If one caught my eye, I read the paper and my mind was blown by what scientists had discovered. I thought – *Wouldn't it be brilliant if I could write about these amazing insights for non-scientist readers?* But which publication would be interested? I could think of none that ran such stories about species from across the world. *The Wire* had recently started a science section. New platforms still finding their feet were probably more open to trying something different. I wasn't wrong.

I have always enjoyed the company of scientists. The column gave me the privilege of being lost in their world, week after week. Most of them were generous with their time, explaining the science and discussing the challenges they faced. There are too many of them to name individually here but you have read their names and work in the preceding pages.

'Think you might compile these pieces into a book one day?' Mukunth asked when I had written forty essays. But it wasn't until a year later, when Mahesh Rangarajan urged me to compile them that I made a move. Mahesh is a former director, Nehru Memorial Museum and Library, New Delhi, and professor of Environmental Studies and History at Ashoka University. He features big time in my writing career, suggesting who to write for, introducing me to editors, and urging editors to publish me. His words of encouragement left me with a swollen head for the rest of the day.

I spent three days compiling the essays I had written until then, turning hyperlinks to footnotes. The editor of one large publishing company said it didn't work for them. I asked a friend, M.D. Madhusudan, a scientist at Nature Conservation Foundation, if he had any ideas. He wrote a glowing introduction to an editor at a foreign university press, the kind of letter that embodies a tremendous amount of warmth, an insight into what editors and writers need, and one that would have taken me days to achieve that degree of finesse. It's no reflection on his efforts that the manuscript didn't find a home there.

I was delighted that Krishan Chopra and Ananya Borgohain adopted the manuscript for the HarperCollins India stable.

Rohan Mathias not only offered the hospitality of his home in the wilds of Masinagudi but also came up with the title for this book.

Janet DePenning and attorneys, Malavika T.V. and Sunil Jose, at DePenning & DePenning advised on intellectual property rights, a subject that turned my grey cells into mush.

Mita Kapur of Siyahi guided me through the legalese of contracts.

Romulus Whitaker, fellow curious cat, is responsible for opening the natural world to me. He is to blame for the past two and half decades of my obsession with the wild.

My parents, Sivakami and K.R. Lenin, still wonder where this strange daughter of theirs came from.

Thank you all. Lastly, thanks to you, dear reader, for staying with me until this page.

About the Author

Janaki Lenin lives in a village near Chennai, Tamil Nadu, with her husband, Romulus Whitaker. Her freelance journalism career began in 2004. Prior to that, she produced wildlife documentaries for National Geographic Television with Rom, and before that, she edited endless soap opera sagas and run-of-the-mill commercials.

Janaki was an indifferent school student, counting down the days to vacation when she could devour books by the dozen. Back then, she didn't suspect she'd eventually become an author herself. Although she graduated in filmmaking, she learned the skill as an apprentice. While researching for wildlife films, she taught herself science by reading journals, referring to dictionaries, and listening to researchers. That changed the way she viewed animals, prompting a passion to share the wonders of nature. This goal didn't change when she switched careers to become a writer.

Janaki served as the Coordinator of the Agumbe Rainforest Research Station from 2005 until 2009. In 2007, she co-